Organic Reaction Mechanism and Applications

Organic Reaction Mechanism and Applications

Zeba N. Siddiqui

AKANSHA PUBLISHING HOUSE
NEW DELHI-110002 (INDIA)

AKANSHA PUBLISHING HOUSE
4649-B/21, Ansari Road
Darya Ganj, New Delhi - 110 002
Ph.: 23263193 / 9811582579 / 09435116718
Fax: 011-23263193
Email: ektabooks@yahoo.com
www.akanshapublishinghouse.com

Organic Reaction Mechanism and Applications

First Published 2009

ISBN 978-81-8370-211-9

PRINTED IN INDIA

Published by Akansha Publishing House, New Delhi and Printed at Mehra Offset, Delhi.

Dedicated to

My mother

Husn Bano Siddiqui

Contents

Preface

The aim of this book is to present a detailed account of organic reaction mechanism and its applications. Efforts have been made to overcome every possible difficulty arising out while writing the mechanism of a particular reaction. The book has been written in a simple and lucid style. It has been divided into twelve chapters. In the first chapter synthesis of aldehydes by various methods has been discussed. In the second and third chapter synthesis of ketones and unsaturated carbonyl compounds have been discussed.

Synthesis of flavonoids, amines and aromatic compounds have been devoted to chapters fourth, fifth and sixth. Synthesis of carboxylic acids, esters, olefins / alkynes and alcohols / phenols are described in seventh, eighth, ninth and tenth chapters. The last two chapters contain synthesis of miscellaneous compounds and molecular rearrangement.

A book of this kind cannot be written originally and thus I do not claim any originality since I have drawn subject matter from various published and unpublished sources of various departments, laboratories and institutes. I thank them all very sincerely.

The book has been written generally for students of B.Sc. (Chemistry & Industrial Chemistry), M.Sc. (Chemistry & Industrial Chemistry), and particularly for research students,

College/University teachers and those who are involved in day to day research work related to organic chemistry.

I wish to place my abiding sense of gratitude to Prof. P.K.Abdul Azis, Vice-Chancellor, .Aligarh Muslim University for his invaluable inspiration and guidance.

It gives me immense pleasure to express my deep appreciation to Prof. V.K.Abdul Jaleel, Registrar, A.M.U.Aligarh for his help and advice while writing this book.

I wish to record my indebtedness to Prof. K.S. Siddequi Chairman, Department of Chemistry for his invaluable suggestions and encouragement.

I express my sincere gratefulness to Prof. Zafar Ahmad Siddiqui for his constant support, cooperation and invaluable suggestions.

My sincere appreciation and heartfelt thanks are due to Prof. S.A. Nabi for his persistent interest, generous cooperation and timely suggestions.

I wish to express words of special thanks and sincere appreciation to my husband, Dr. Nafees A. Khan, Reader, Deptt. of Commerce who inspired me at all stages of the work.

Lastly, I am thankful to my .sons Waqas, Wasiq and Wamiq for keeping patience during the preparation of this manuscript.

Finally I wish to thank M/s Akansha Publishing House for their admirable cooperation.

Dr. Zeba N. Siddiqui

1

Synthesis of Aldehydes

1.1. Akabori amino acid reaction

Aldehydes **2** are obtained when α amino acids **1** are heated with sugars.[1]

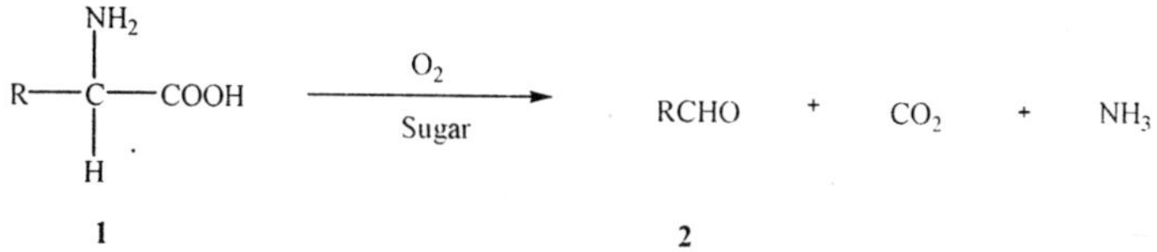

The reaction occurs as a result of oxidative decomposition of α amino acid. However, reduction of α amino acid/esters **3** to α amino aldehyde **4** takes place in the presence of sodium amalgam and hydrogen chloride.

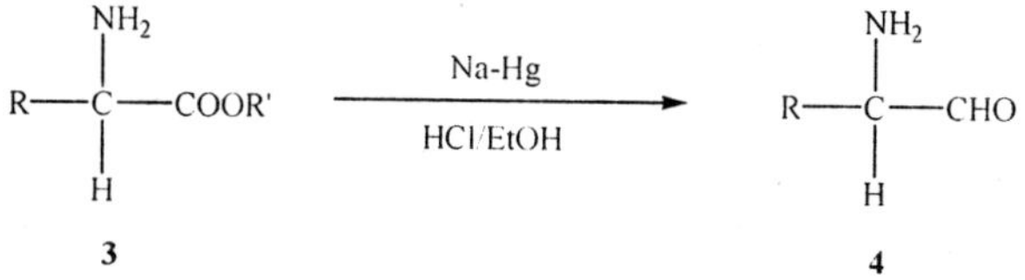

$R' = H, C_2H_5$

1.2. Bouveault Aldehyde synthesis

The reaction of N, N-disubstituted formamide **5** with Grignard reagent or organolithium compound to give homologous aldehyde **6** is called as Bouveault- Aldehyde reaction.[2]

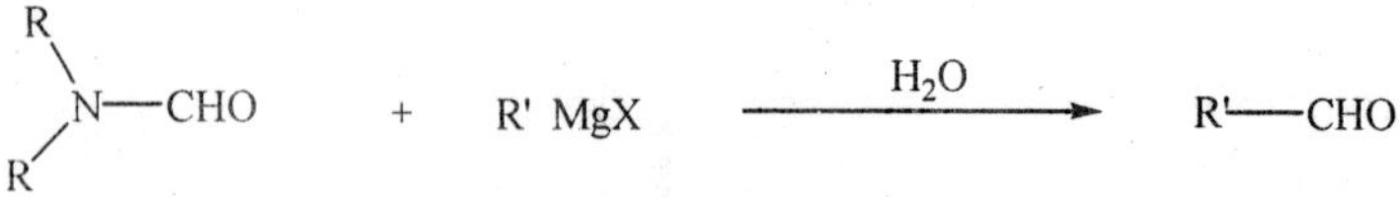

5 **6**

The mechanism involves attack of Grignard reagent on carbonyl group of formamide **5** to give addition product **7** which is hydrolysed to give **6**.

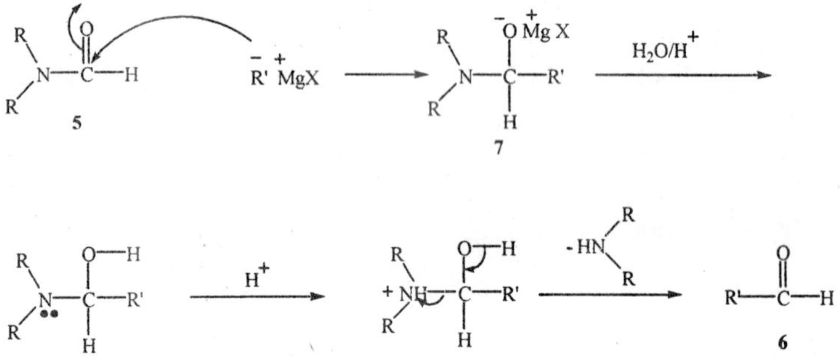

1.3. Reimer Tiemann reaction

The reaction involves synthesis of aromatic aldehyde **8** from phenol, chloroform and alcoholic solution of alkali.[3,4]

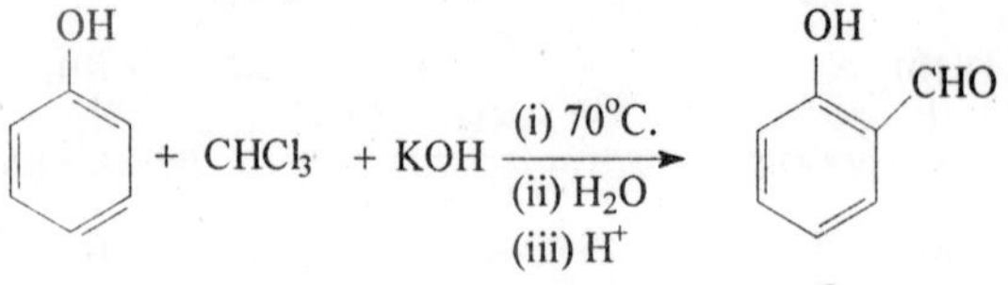

8

The mechanism involves following steps:

1. Formation of electrophile, dichlorocarbene **9** by the reaction of chloroform and alkali.

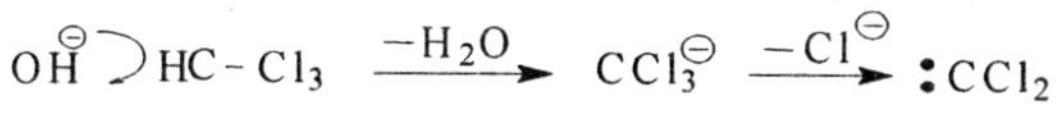

9

2. Phenol reacts with electrophile **9** (electrophilic substitution) to form the product **10**.

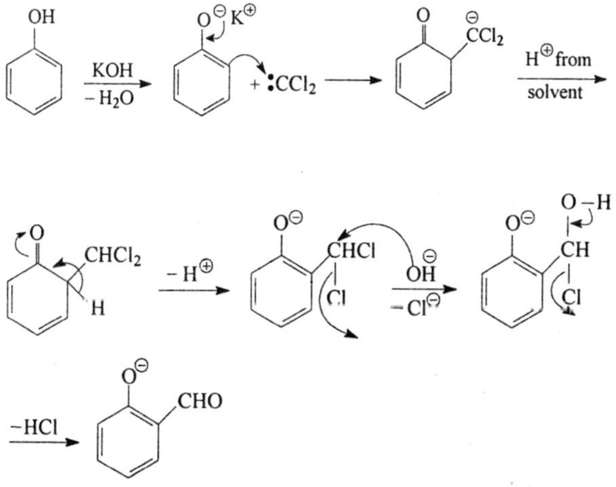

10

3. Free salicyldehyde **8** is obtained by acidification of **10** with HCl.

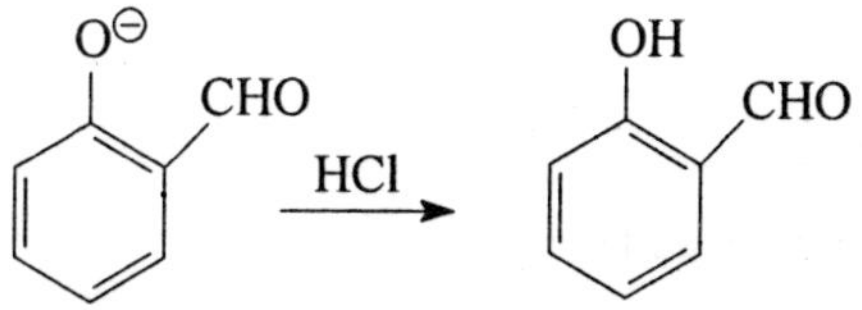

10 **8**

Chloroform is preferred over carbon tetrachloride and dichloromethane because chloroform being polar in nature undergoes alkaline hydrolysis faster. Though the hydrolysis reaction is reversible, the product formation in forward direction takes place by added chloride ions.

Synthetic utility

1. 5-Formyl barbituric acid **12** is obtained by stirring a mixture of barbituric acid **11,** chloroform and alcoholic potassium hydroxide followed by acidic hydrolysis.[5]

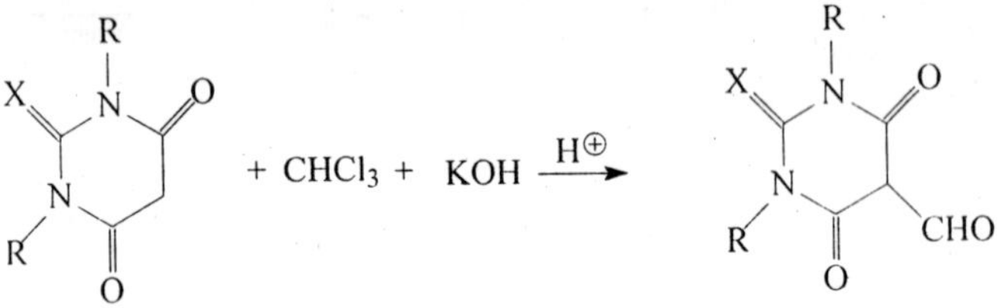

11 **12**

R = H, Me

X = O, S

2. 2-Formyl pyrrole and 2-formyl thiophene have been synthesized successfully using a mixture of chloroform and alkali.

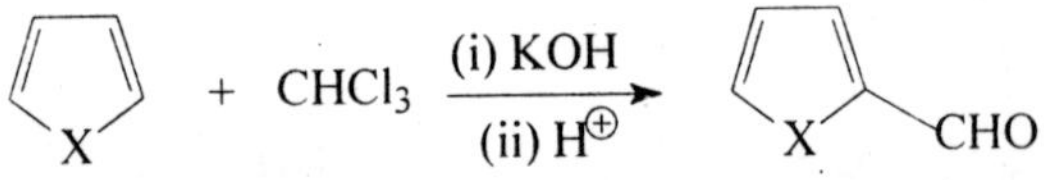

X = NH, S.

1.4. Sommlet reaction

The reaction is employed for the synthesis of aldehydes. The reaction involves treatment of aralkyl halides **13** with hexamethylene tetramine **14** to give quaternary ammonium salt **15** which after hydrolysis yields aldehydes.[6]

$$ArCH_2-X + C_6H_{12}N_4 \xrightarrow[\text{reaction}]{SN^2} Ar-CH_2-C_6H_{12}N_4^{\oplus}\, X^{\ominus} \xrightarrow{\Delta / H_2O} ArCHO$$

13   14   15

Suck-Ju Hong,[7] reported the synthesis of dialdehyde **16** from naphthalene by Sommlet reaction.

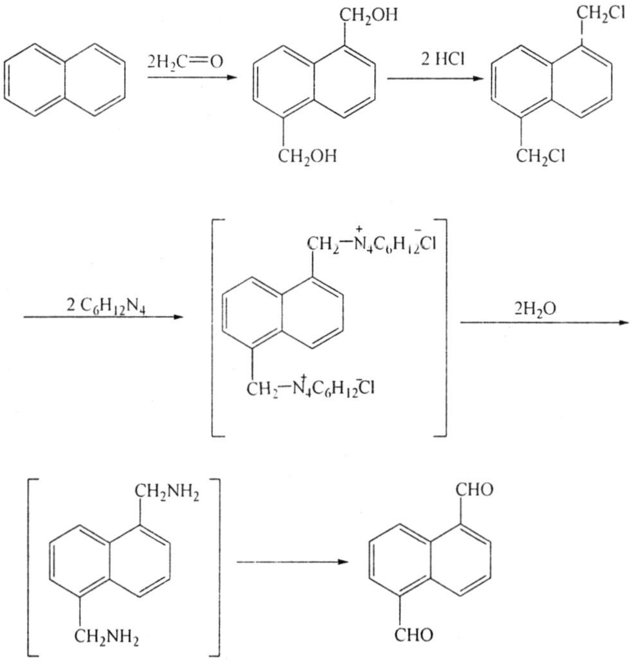

16

1.5. Sonn-Muller reaction

The reaction involves formation of imidoylchloride **18** as intermediate from aromatic anilide **17** and phosphorous pentachloride. The intermediate **18** is reduced to imine **19** upon treatment with stannous chloride and hydrochloric acid. Hydrolysis of imine finally gives aldehyde **20**.[8]

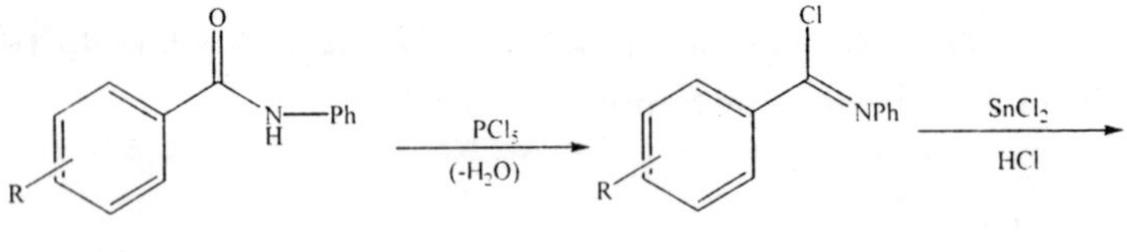

17 **18**

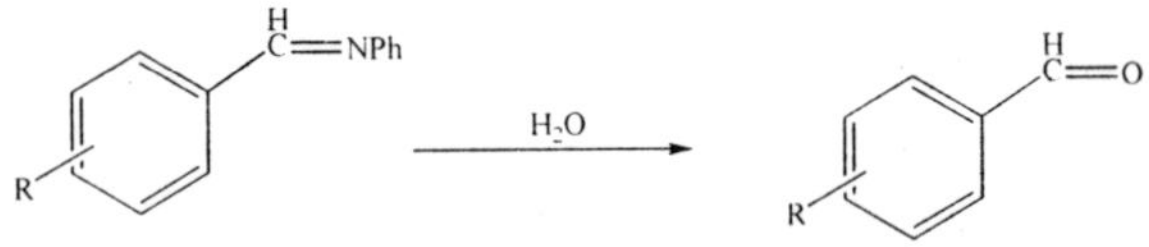

19 **20**

1.6. Stephen aldehyde synthesis

The reaction is used for the synthesis of aromatic aldehydes from nitriles (cyanides). The reaction occurs by treatment of the nitrile **21** with stannous chloride in the presence of hydrochloric acid. The intermediate imine salt **22** then hydrolyses to aldehyde.[9]

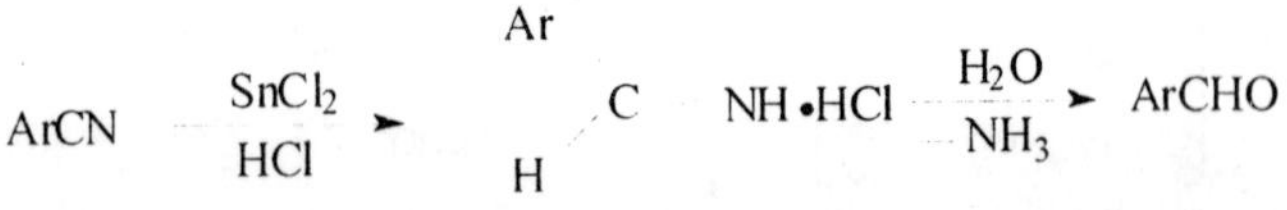

21 **22**

1.7. Vilsmeir Haak reaction

The reaction is used for the formylation of activated aromatic or heteroaromatic compounds using Vilsmeir reagent (disubstituted formamide + phosphorous oxychloride).[10] The reaction involves following steps:

1. N,N-Dimethylformamide forms reactive species, iminium salt **23** upon treatment with phosphours oxychloride.

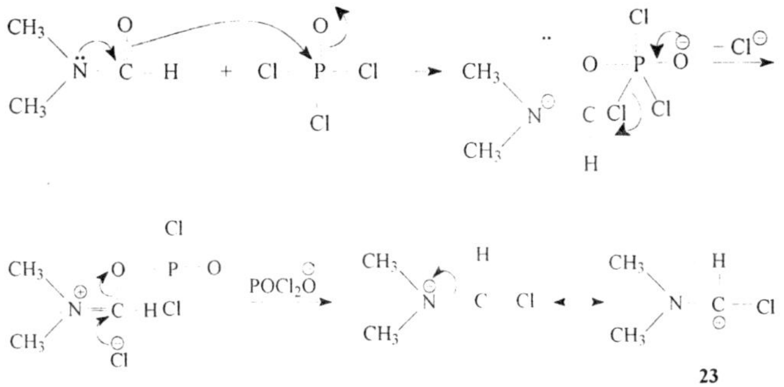

23

2. The iminium salt **23** acts as electrophile and reacts with aromatic compound through electrophilic substitution reaction to give **24**. The intermediate salt **24** upon hydrolysis forms the aldehyde **25**.

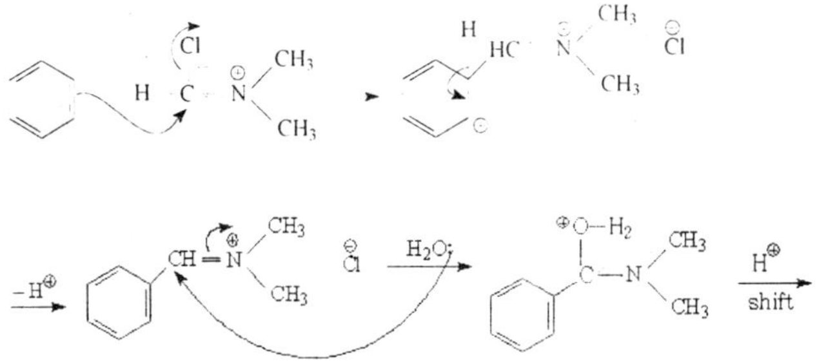

24

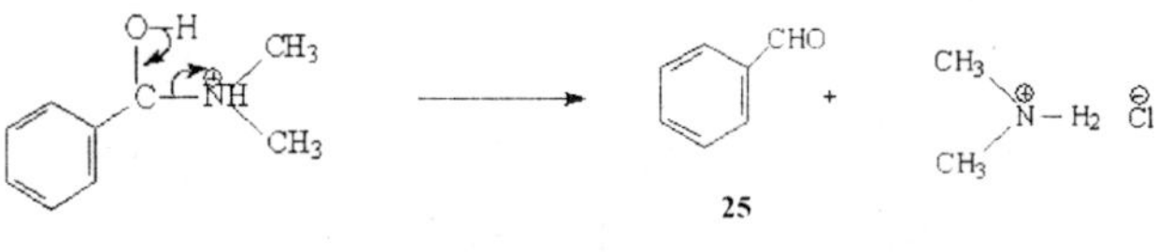

25

Thus, some of the commonly used aldehydes have been synthesized successfully through Vilsmeir Haak reaction.

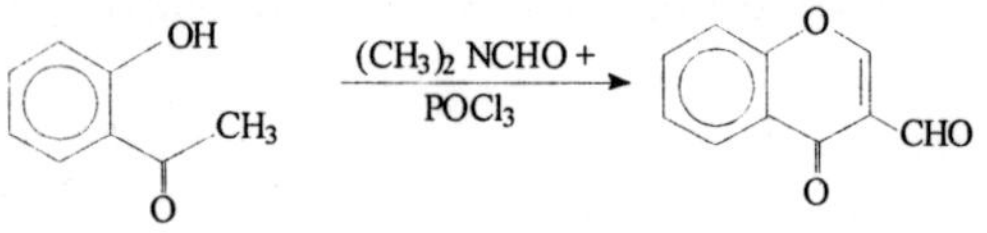

(Ref 11)

2.

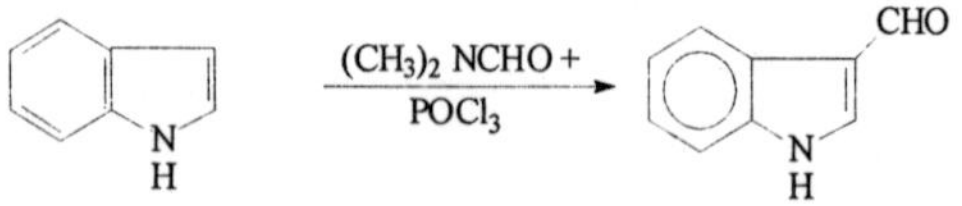

(Ref12)

3.

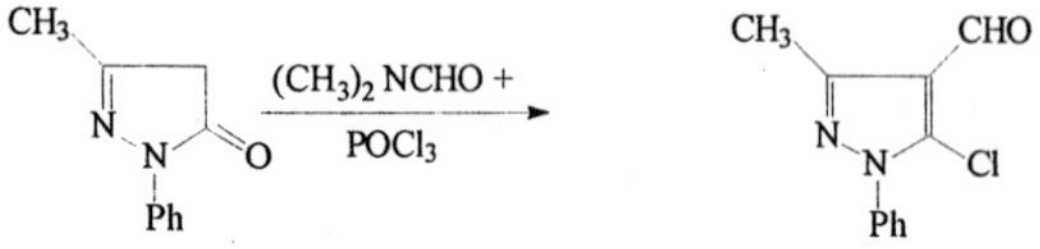

(Ref 13)

REFERENCES

1 E. Takagi et al, *J. Pharm. Soc. Japan*, 1951, **71**, 648; 1952, **72**, 812.

2 L. Bouveault, *Bull. Soc. Chim. France*, **1904**, **31**, 1306 and 1322.

3 K. Reimer, F. Tiemann, *Ber.*,1876**, 9,** 824, 1268, 1285.

4 H. Wynberg, *Chem. Rev.*,1960, **60**, 169 .

5 B. S. Jursic, D. M. Neumann, *Tetrahedron Lett.*, 2001, **42 (48)**, 8435.

6 M. Sommlet, *Compt. Rend.* ,1913, **157**, 852.

7 Suck-Ju Hong, *J. Korean Chem. Soc.*, 1972, **16** (2).

8 A. Sonn and E. Muller. *Ber.*, 1919, **52,** 1927.

9 H. Stephen, *J. Chem. Soc.*, 1925**, 121**, 1814.

10 A. Vilsmeir and A. Haak, *Ber.*, 1927, **60,** 119.

11 A. Nohara, T. Umetani and Y. Sanno, *Tetrahedron*, 1974, **30,** 3553.

12 G. F Smith, *J. Chem. Soc.,* 1954, 3842.

13 J. Becher, P.H. Olesen, N.A. Knudsen and H. Toftlund, *Sulfur Lett.* 1986, 4, 175.

2

Synthesis of Ketones

2.1. Acyloin condensation

Two molecules of an ester or a diester when treated with sodium forms α-hydroxy ketone (acyloin).

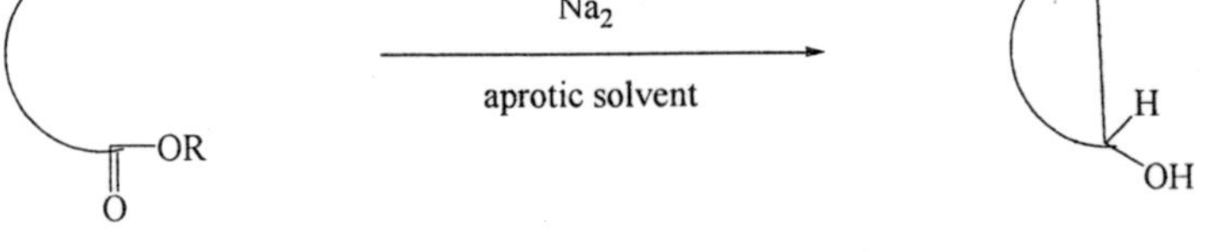

This reaction is known as acyloin condensation and usually carried out in an aprotic solvent such as ether or high boiling hydrocarbon such as trimethyl chlorosilane etc.[1]

The reaction involves reductive coupling of esters by sodium as shown in following steps.

1. Reduction of carbonyl group of ester **1** by sodium to form radical anion **2**.

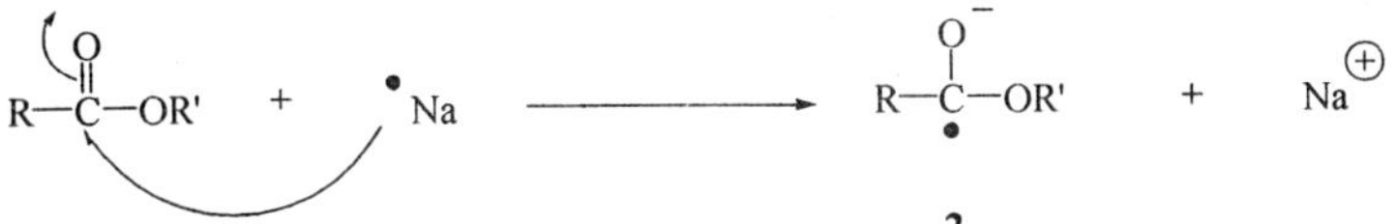

1 2

2. Coupling of two molecules of radical anions **2** to form α diketone **3**.

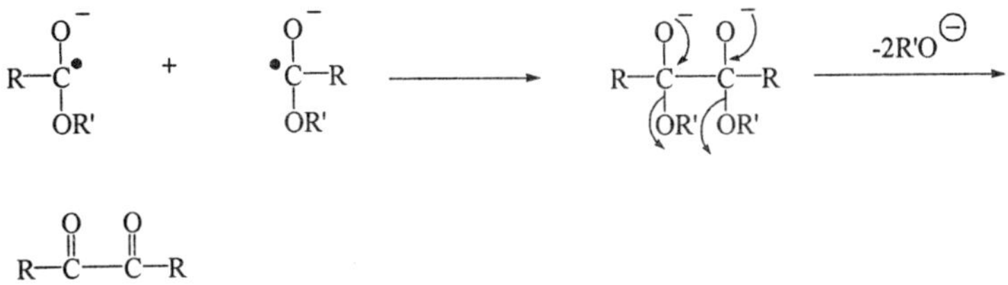

3

3. Reduction of two carbonyl groups of **3** by sodium to form the product acyloin **4**.

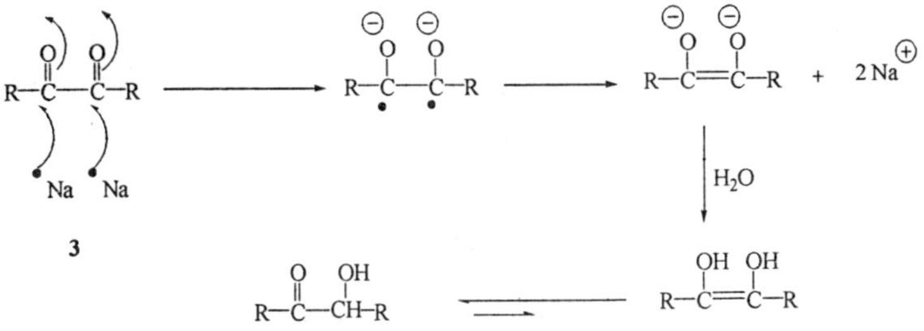

4

The acyloin condensation is a good method for the synthesis of small, and large carbocyclic compounds.

Synthetic applications

1. *Synthesis of small and normal ring containing acyloins.*

A four membered acyloin **6** was synthesized from diester **5**.[2]

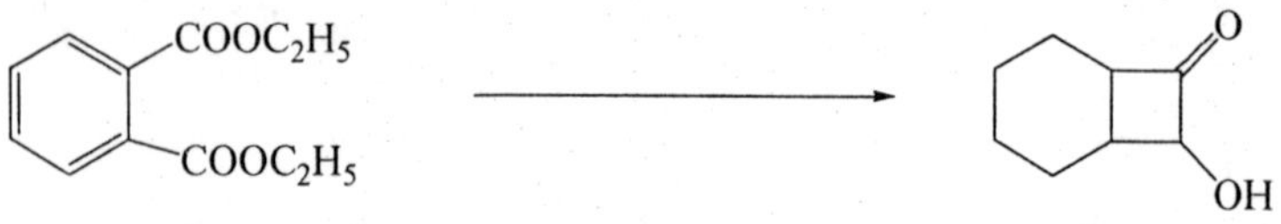

5 **6**

Synthesis of another six membered acyloin adipoin **7** has been reported from diester **5a**.[3]

5a **7**

Solvents generally used are benzene, toluene, liquid ammonia, ether and xylene.

2. *Synthesis of medium ring containing acyloins.*

Acyloin condensation reaction has been also used for the synthesis of medium size ring containing compounds (8-11 membered) e. g.[4]

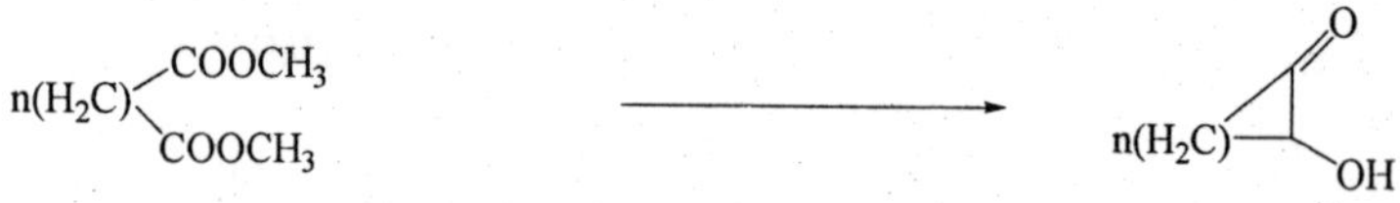

Various substituted medium size ring containing acyloins have been prepared successfully e.g.[5]

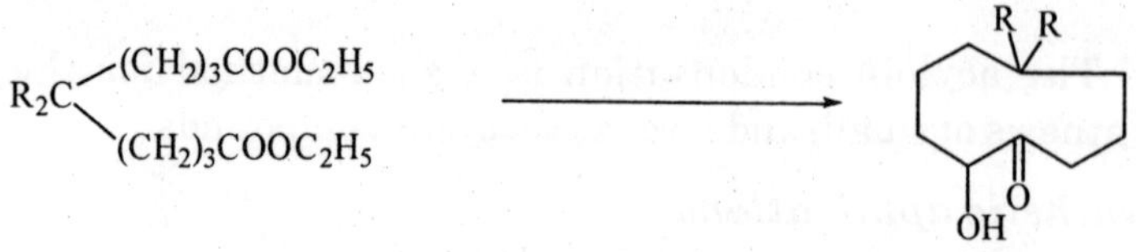

R= CH_3, C_2H_5

Synthesis of medium ring compounds containing a heteroatom is also reported.[6]

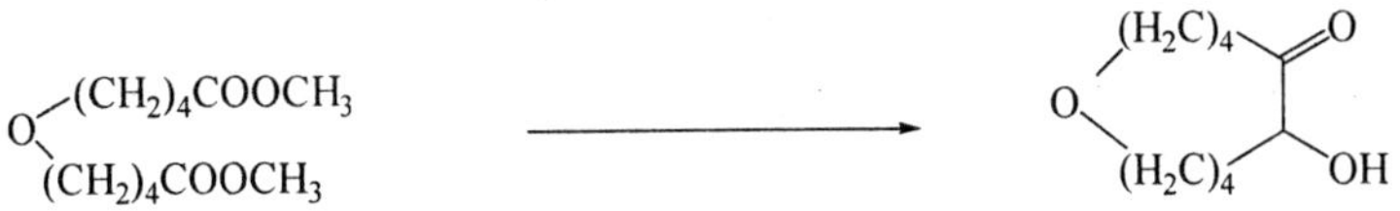

3. *Synthesis of large ring containing acyloins.*

The large ring containing acyloins have been synthesized by using a diester containing a triple bond.[7,8]

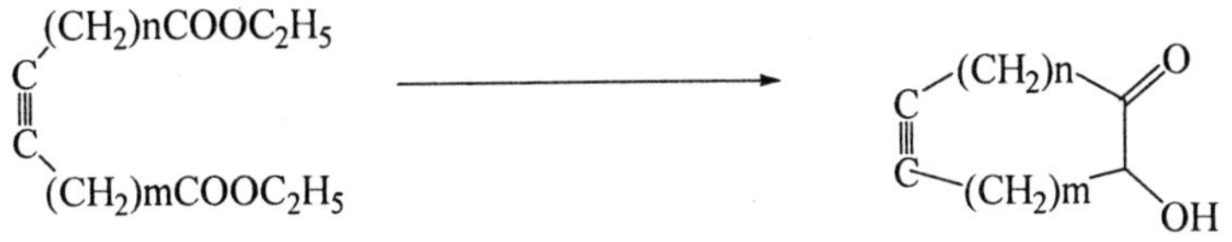

n= 5, m= 9 or n=m=7

Steroidal acyloins **9** have been also used as intermediate for functionalization.[9]

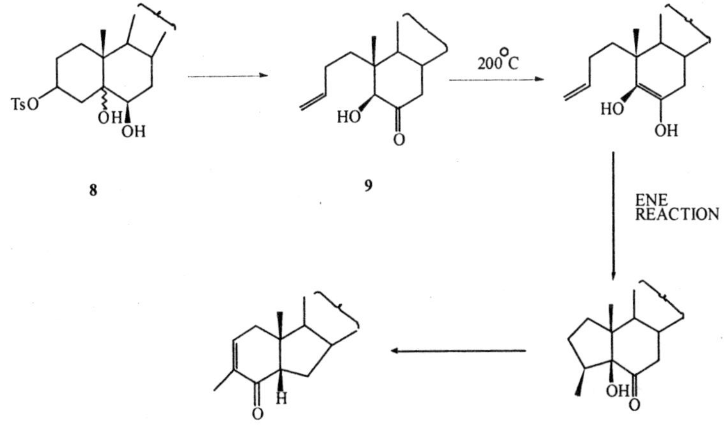

The intermediate acyloin is obtained from **8** as shown below:

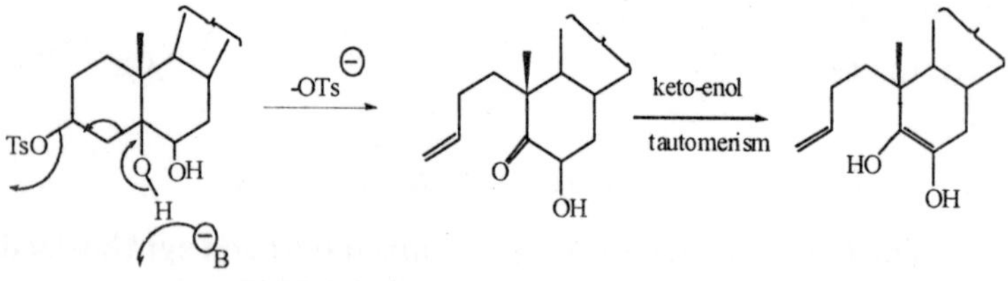

8 **9**

2.2. Benzoin condensation

The self condensation reaction of aromatic aldehyde specifically catalysed by cyanide ion to give α-hydroxyketone **10** is called as benzoin condensation.[10] The α-hydroxyketone **10** is also known as benzoin. The benzoin condensation reactions is applicable to aromatic aldehydes containing no α-hydrogen and not to aliphatic aldehydes.

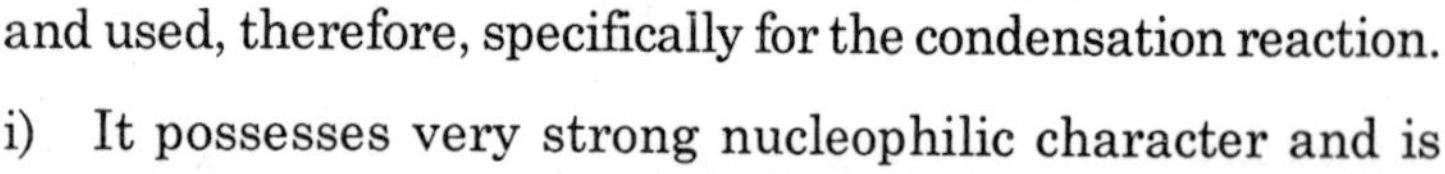

10

The catalyst cyanide ion has following special features and used, therefore, specifically for the condensation reaction.

i) It possesses very strong nucleophilic character and is added very easily to the carbonyl group.

ii) Due to powerful electron withdrawing nature of cyanide ion it exerts negative inductive effect (-I effect) on carbon and as a result of this C-CN bond becomes polar creating a partial positive charge on carbonyl carbon and partial negative charge on cyanide group. Due to presence of partial positive charge on carbon electron density of C-H bond gets disturbed and C-H bond also becomes polar creating a slightly positive charge on hydrogen. Thus, due to unequal sharing of electron pair between C-H bond aldehydic hydrogen in the form of proton is easily shifted to negatively charged oxygen to form carbanion **12** which is stabilized by benzene ring as well as by cyanide group.

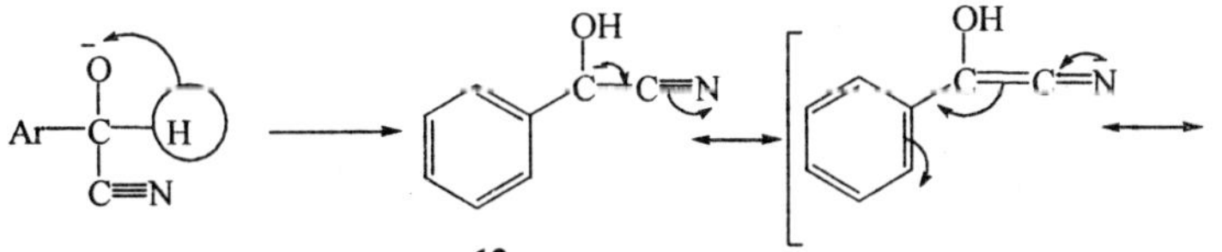

12

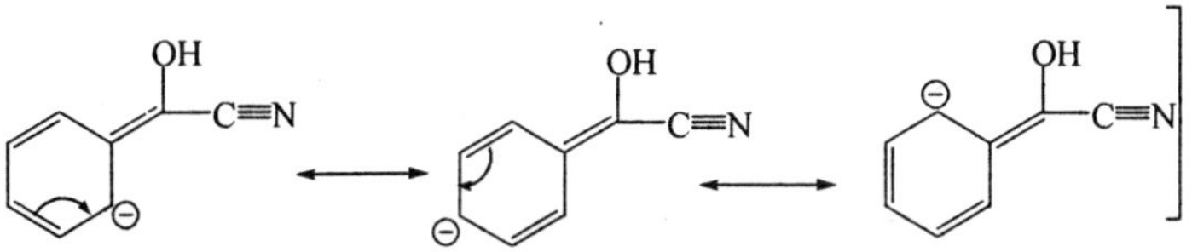

iii) Cyanide group is a good leaving group and can depart from cyanohydrin in the intermediate **13** to give the final product **10**.

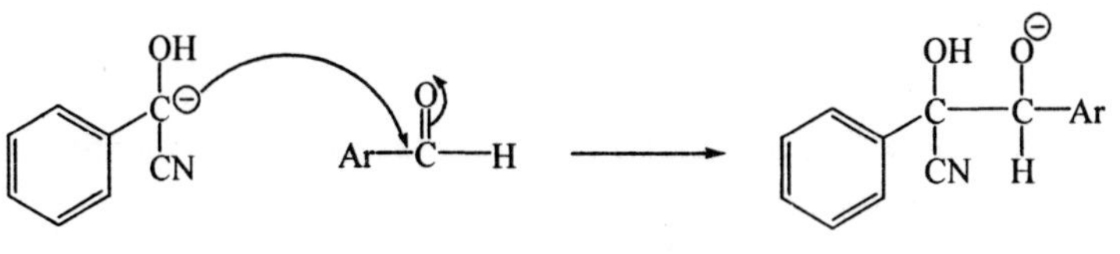

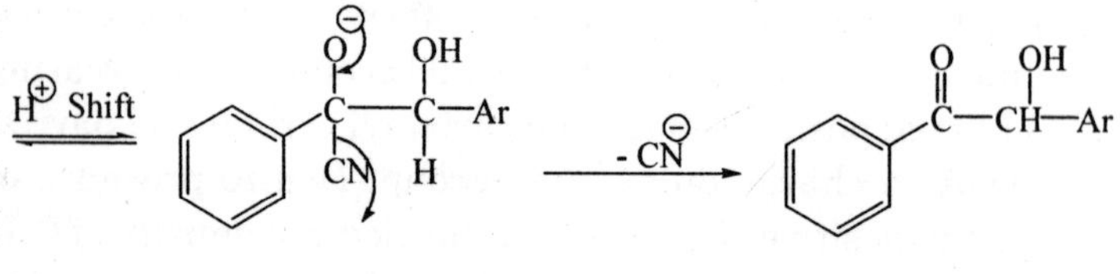

13 10

2.3. Houben-Hoesch Reaction

The reaction is used for the synthesis of o-hydroxy ketones from phenols. The product **15** is obtained when phenol **14** is treated with alkyl nitrile in the presence of hydrochloric acid and aluminium chloride or zinc chloride as Lewis acid catalyst.[11,12]

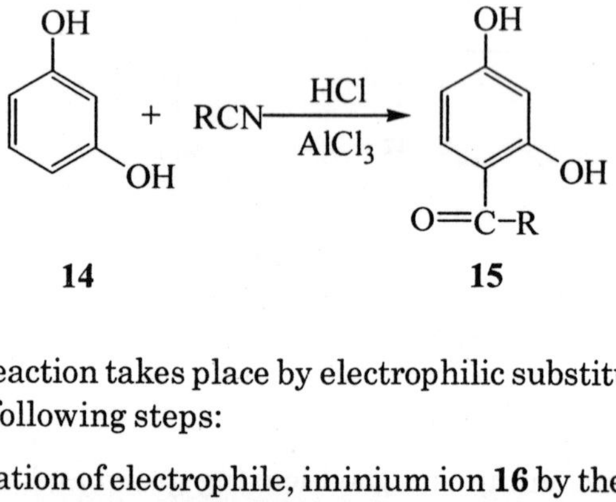

14 15

The reaction takes place by electrophilic substitution and involves following steps:

1. Formation of electrophile, iminium ion **16** by the reaction of alkyl cyanide and Lewis acid.

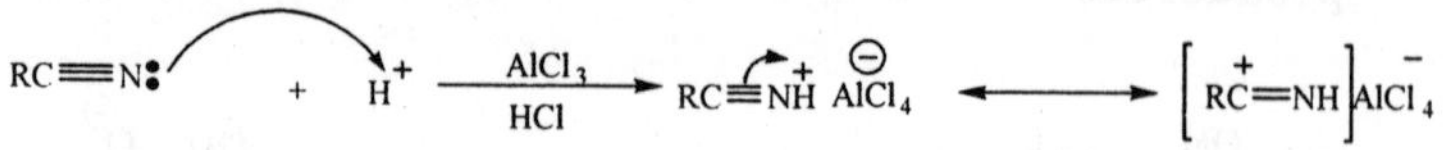

16

1. Electrophilic attack by benzene takes place to form the intermediate **17**.

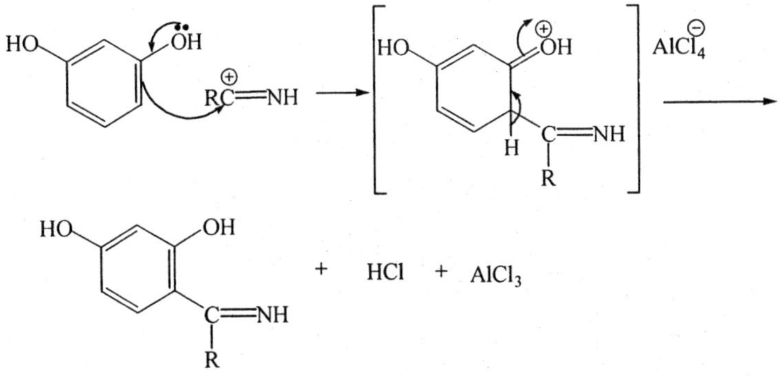

17

Formation of product **15** takes place by hydrolysis of intermediate **17**.

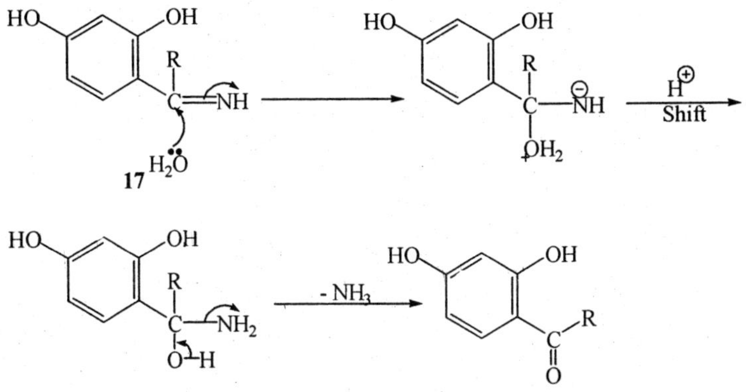

15

The Houben- Hoesch reaction is an extension of Gattermann synthesis. The reaction is more successful with phenols carrying more than one hydroxyl groups or electron releasing groups such as alkoxy etc.

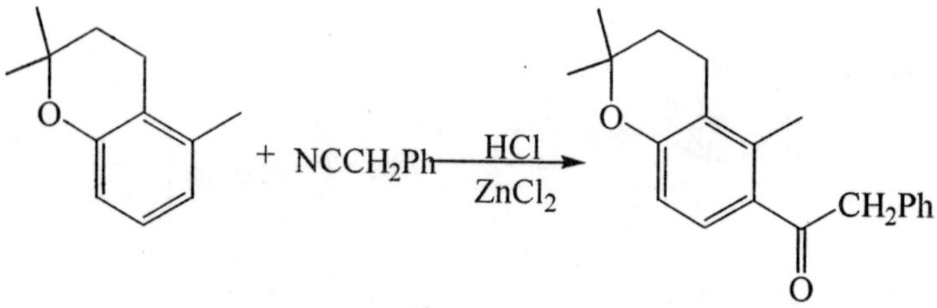

2.4. Nef reaction

Conversion of primary / secondary nitroalkanes to aldehydes/ ketones upon treatment with acid/ base is called as Nef reaction.

$$CH_3\text{—}\overset{H_2}{C}\text{—}NO_2 \xrightarrow[2.\ H_2SO_4]{1.\ NaOH} CH_3CHO$$

Primary and secondary nitroalkanes are acidic in nature and form salt upon treatment with base. The acidic character is due to – I effect of the nitro group and carbanion **18** formed is stabilized by nitro group through resonance.

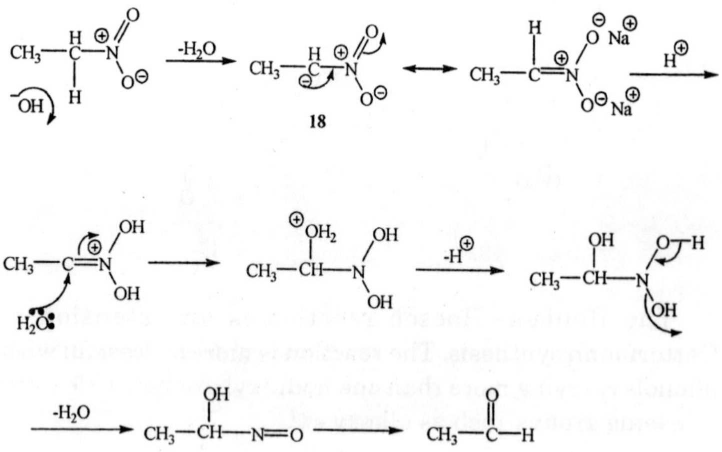

Synthetic applications

(i) *Oxone promoted Nef reaction.* A mild and convenient oxidative Nef reaction in the presence of oxone has been reported. Thus, primary and secondary nitroalkanes generate carboxylic acid and ketones in good yield.[13]

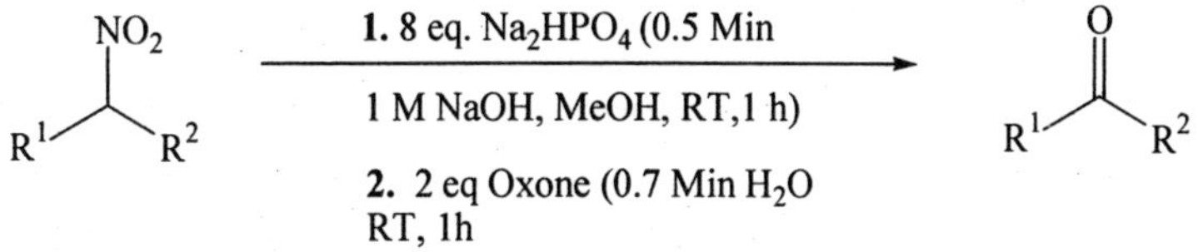

(ii) *Nef reaction of secondary nitro alkanes catalysed by DBU*

Secondary nitroalkanes **19** have been converted to corresponding ketones **20** under basic conditions using DBU in acetonitrile.[14]

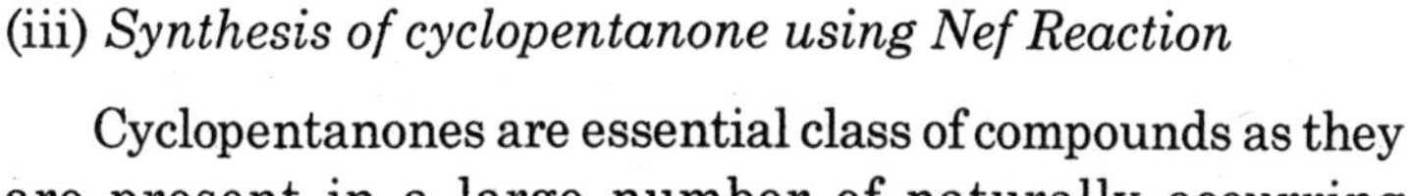

19 **20**

(iii) *Synthesis of cyclopentanone using Nef Reaction*

Cyclopentanones are essential class of compounds as they are present in a large number of naturally occurring compounds. Recently[15] its synthesis has been reported involving conjugate addition of nitroalkanes to α,β-unsaturated ketones to give **21** which is then converted to γ-diketone **22** using Nef reaction. Intramolecular cyclization of **22** ultimately forms cyclopentanone **23**.

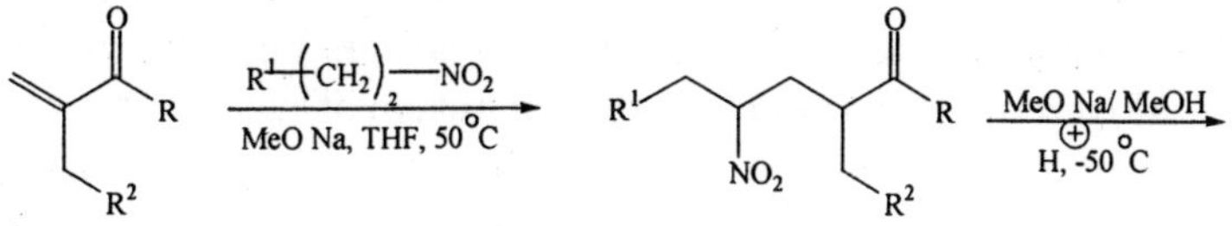

21

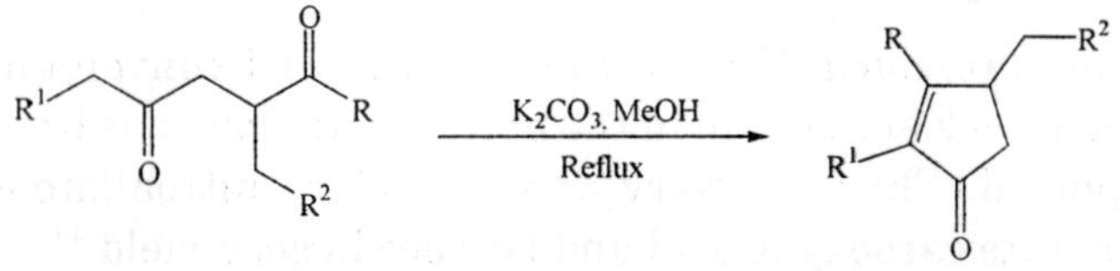

22 **23**

(iv) Aliphatic nitro compounds have been used for the synthesis of pyrazoles through Nef Reaction.

Thus, 1,3- dinitro compounds can be treated as synthetic equivalents for 1,3-diketones through Nef Reaction. Cyclization of 1,3-dinitro compounds to pyrazoles then takes place upon treatment with hydrazines.[16]

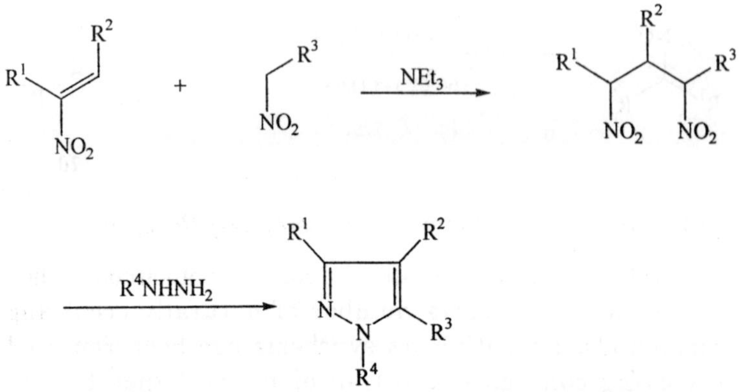

2.5. Nazarov cyclisation reaction

The reaction is used for the synthesis of cyclopentenone from a ketone or acid chloride derivative. The reaction is catalyzed by Lewis acids. Thus, formation of cyclopentenone **25** from divinyl ketone **24** is catalysed by phosphoric acid/ formic acid mixture.[17]

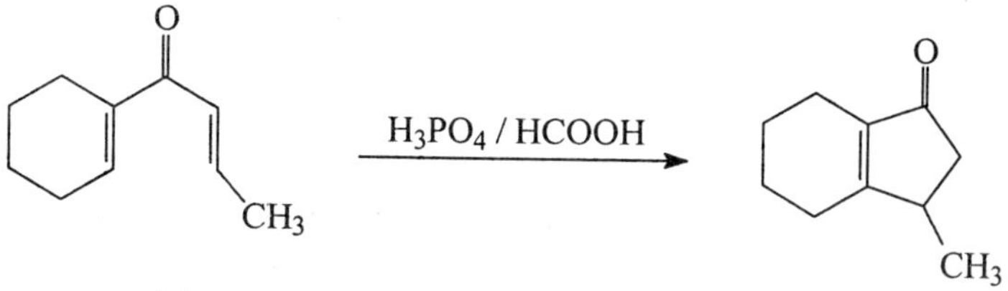

24 25

Cyclization also occurs in the presence of trimethyl silyl alkene.[18]

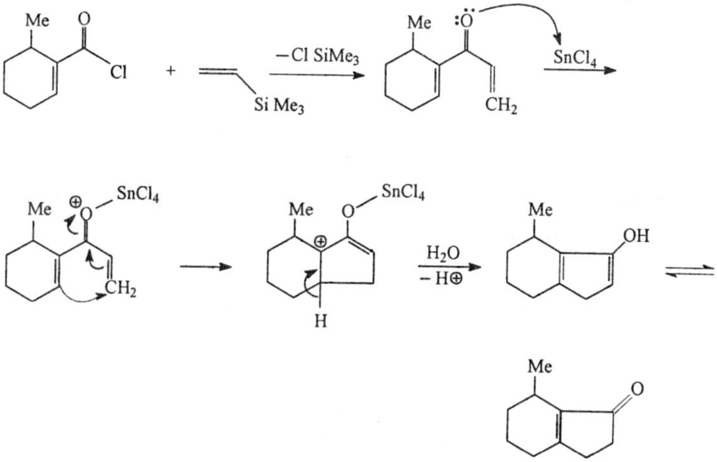

2.6. Norrish type cleavage

This is a photochemical reaction. It is of two kinds, namely, Norrish type I reaction and Norrish type II reaction.[19-22]

Norrish type-I reaction

This reaction involves a cleavage of C-C bond of ketones in a homolytic manner to form acyl free radical and alkyl/ aryl free radical.

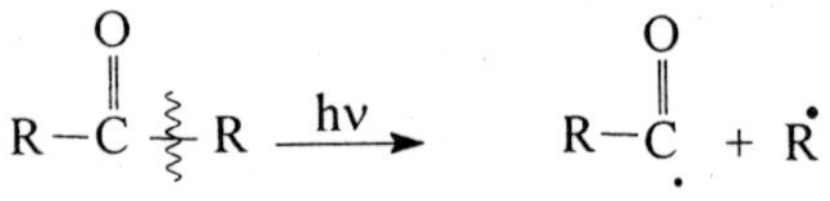

At higher temperature the primary reaction is followed by secondary process which involves decarbonylation of acyl free radical to form an alkane.

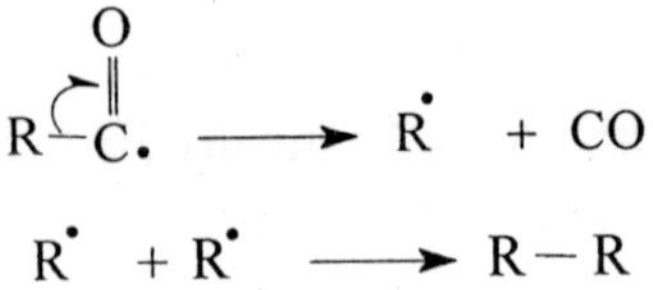

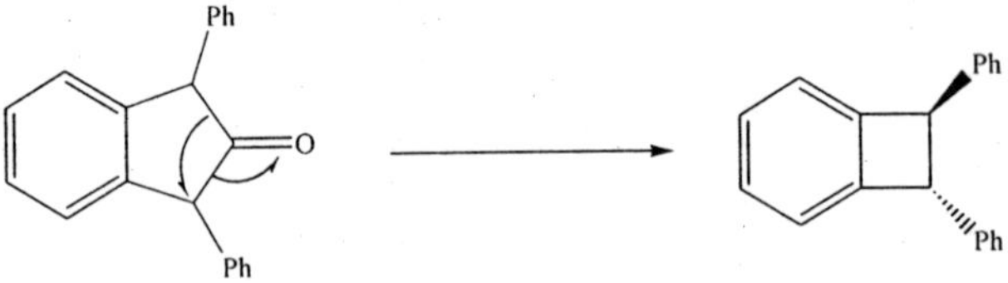

When ketone is unsymmetrical the cleavage takes place in such a manner so as to produce more stable free radicals e.g. tertiary butyl methyl ketone **26** undergoes cleavage to form tertiary butyl radical **28** along with acetyl radical **27** and not methyl **29** and acyl radicals **30**.

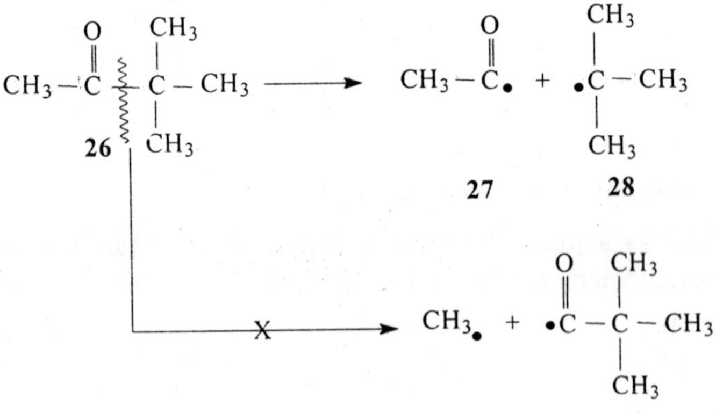

29 **30**

During the reaction carbonyl group undergoes n→π* transition state and the excitation energy of the lowest triplet state of ketone is close to the energy required for breaking of a C - C bond.

Norrish Type II reaction

In this type of reaction aldehydes and ketones containing γ-hydrogen undergo intramolecular γ-hydrogen abstraction to form biradical which undergoes cleavage to give an alkene **31** and an enol **32** or a cyclic hydroxyl compound **33**.

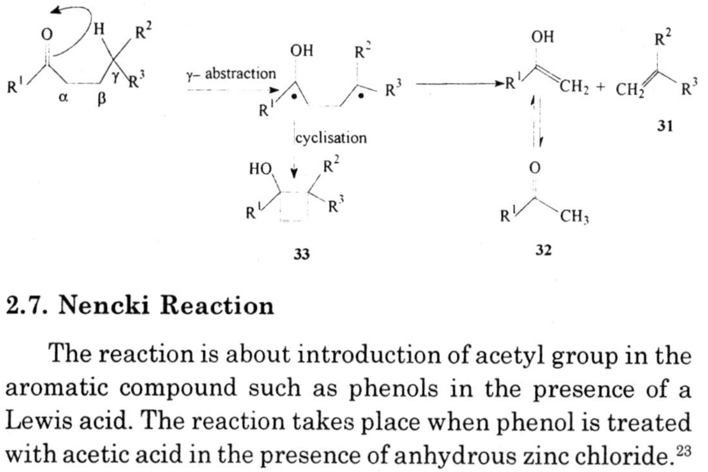

2.7. Nencki Reaction

The reaction is about introduction of acetyl group in the aromatic compound such as phenols in the presence of a Lewis acid. The reaction takes place when phenol is treated with acetic acid in the presence of anhydrous zinc chloride.[23] Polyhydric phenols give good yield than monohydric phenols. Thus, e.g. resorcinol **34** gives resacetophenone **35** in quantitative yield.

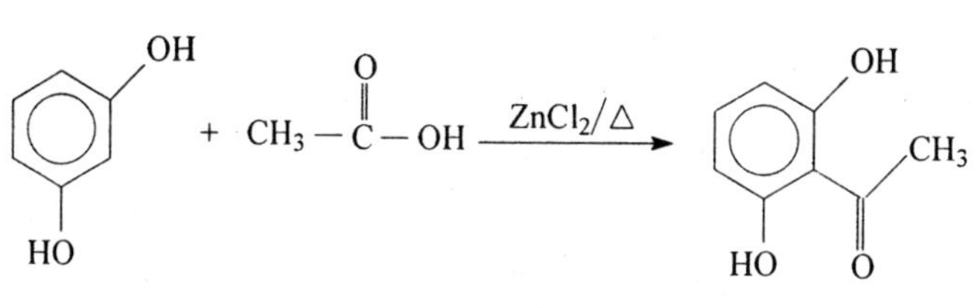

34 **35**

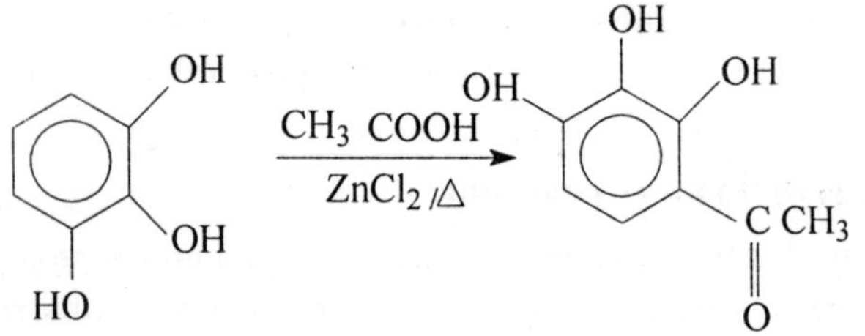

2.8. Tiffeneau-Demjanov Reaction

The reaction involves conversion of β-hydroxy amine **36** to ketone **37a-b** with ratio 2:1, when treated with sodium nitrite and sulfuric acid at low temperature.[24,25]

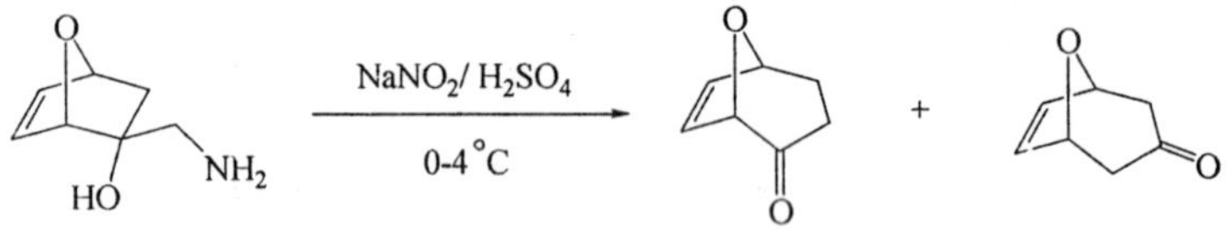

36 **37a** **37b**

2.9.Thorpe-Ziegler Reaction

The reaction is used for the synthesis of cyclic ketones. The reaction takes place by cyclization of dinitriles **38** at high dilution in dialkyl ether in the presence of ether soluble metal alkylanilide and hydrolysis of the resultant iminonitrile with the formation of macrocylic ketones **39**.[26] The reaction takes place as shown below:

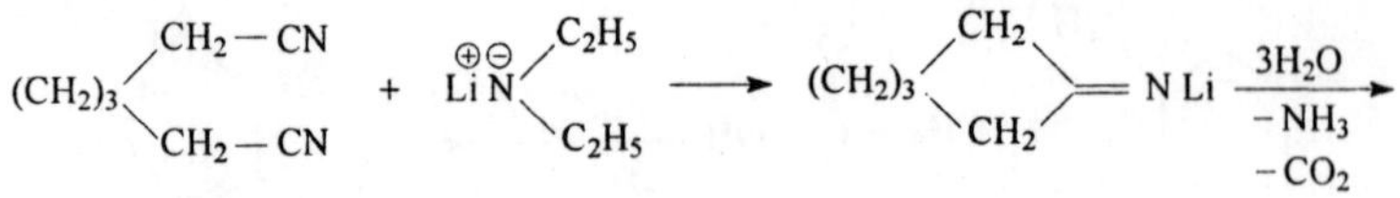

38

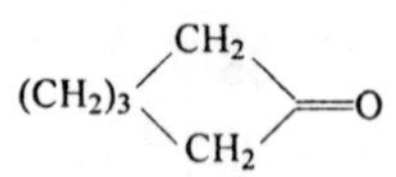

39

The reaction takes place as shown below.

1. Formation of carbanion **40** from CH_2 group which is present adjacent to nitrile group.

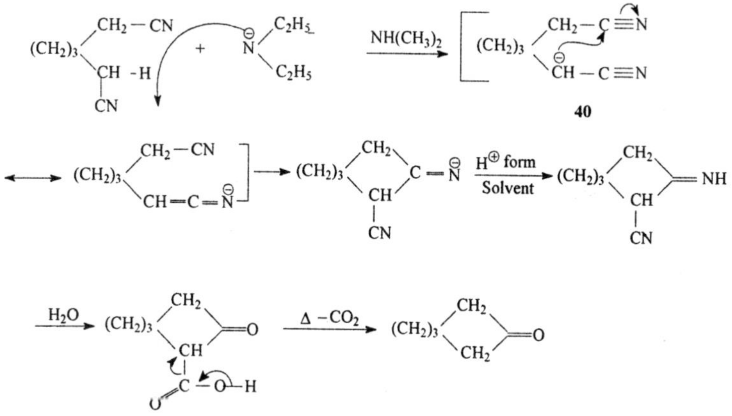

40

Synthetic utility:

The reaction has been used for the synthesis of cyclic ketones **42** from aminonitrile **41.**[27]

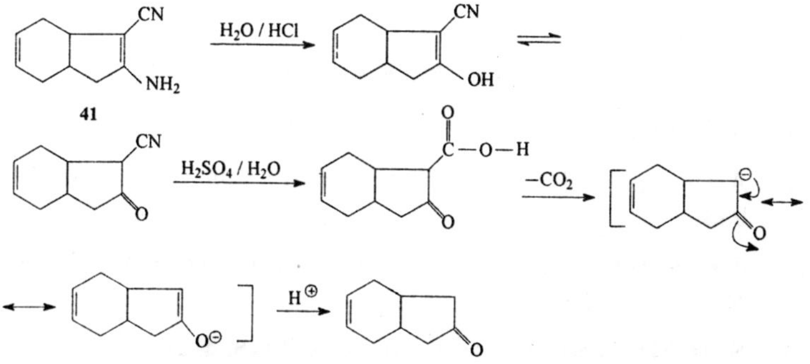

41

42

2.10. Neber Reaction

The reaction involves synthesis of α amino ketones **45**

by treatment of sulfonic esters of ketoximes **43** with potassium ethoxide [28].The reaction proceeds via an azirine intermediate **44**.

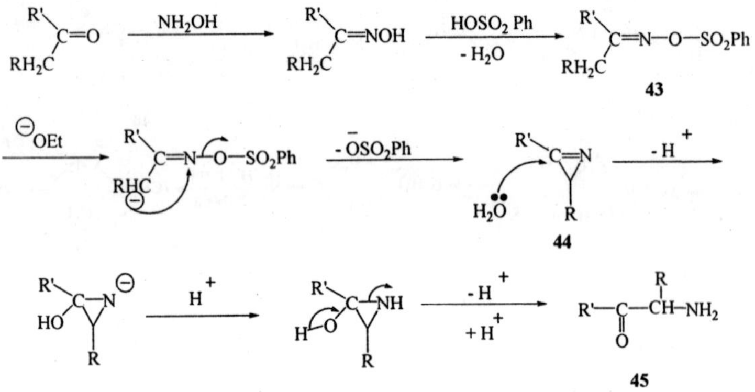

Sometimes the intermediate azirine is obtained as product rather than the α amino ketone.[29]

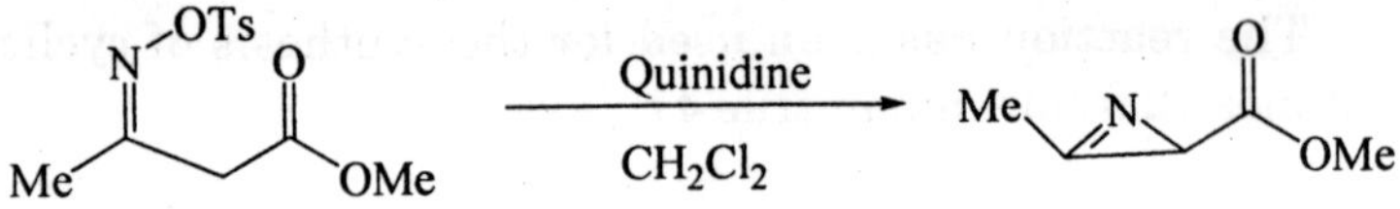

REFERENCES

1. K. T. Finley, *Chem. Review,*1964, **64** (64), 573.
2. A. C. Cope and E. C. Merrick, *J. Am. Chem. Soc.*, 1950, 72, 983.
3. J. C. Sheehan, R. C. O' Neill and M. A. White, *J. Am. Chem. Soc.,* 1950,72, 3376.
4. S. M. McElvain, *Organic reactions,* 1948, 4, 262.
5. A. T. Blomquist, E. S. Wheeler and Y. Chu, *J. Am. Chem. Soc.,* 1955 ,77, 6307.
6. V. Prelog, M.F. El-Neweihy and O. Hofliger, *Helv. Chim. Acta,* 1937,33, 1950.

7. R. Epsztein and I. Marszak, *Compt. Rend*, 1956, **243**, 283.

8. I. Marszak, J. Guermont and R. Epsztein, *Bull. Soc. Chim. France*, 1960, 1807.

9. B. A. Marples and C. D. Spilling, *Tetrahedron*, 1992, **48**, 4017.

10. A. J. Lapworth, *J. Chem. Soc.* , 1903, **83**, 995; 1904, **85**, 1206.

11. K. Hoesch, *Ber.*, 1915, **48**, 1122.

12. J. Houben, *Ber.*, 1926, **59**, 2878

13. P. Ceccherelli, M. Curini, M. C. Marcotullio, F. Epifano, and O. Rosati, *Synth. Commun.*, 1998, **28**, 3057.

14. R. Ballini, G. Bosica, D. Feorini and M. Petrini, *Tetrahedron Lett.*, 2002, **43 (30),** 5233.

15. N. Lahmar, T. B. Ayed, M. Bellassoued and H. Amri, *Beilstein J. Org. Chem.*, 2005, *1* (II); DOI: 10:1186/1860-5397-1-11.

16. F.C. Escribano, M. P. D. Alcantara and A. Gomez-Sanchez, *Tetrahedron Lett.* , 1988, **29 (46),** 6001.

17. I.N. Nazarov, I.B. Torgov and L.N. Terekhova, *Izv; Akad, Nauk. SSSR Otd. Khim. Nauk,* 1942, 200.

18. Y. Sugimoto, S.C.M. Wigchert, J.W.J.F. Thuring and B. Zwanenburg, *J. Org. Chem.*, 1998, **63**, 1259.

19. R.G. W. Norrish, C.H. Bamford, *Nature*, 1936**, 138**, 1016; 1937, **140**, 195.

20. J.D. Coyle, H.A. J. Carless, *Chem. Soc. Rev.* ,1972**, 1**, 465.

21. P.J Wagner, *Acct. Chem. Res.* 1971, **4**, 168.

22. R.P. Wayne, *Principles and Applications of Photochemistry*, Oxford Univ. Press: Oxford, 1988, p. 53.

23. M. Nencki, N. Sieber, *J. Prakt. Chem.* ,1881, **23**, 147.

24. M. Tiffeneau, et al., *Compt. Rend,* 1937, 205, 54.

25. D. Fattori, S. Henry and P. Vogel, *Tetrahedron* ,1993, **49**, 1649.

26. K. Ziegler et al., *Ann.* , 1993, 504, 94.

27. J.J. Bloomfield and P.V. Fennessey, *Tetrahedron Lett.*, 1964, 2273.

28. P. W. Neber and A. Burghard, *Justus Liebigs Ann. Chem.* ,1932, **493**, 281.

29. M. M. H. Verstappen, G. J. A. Ariaans and B. Zwanenburg, *J. Am. Chem. Soc.*, 1996, **118**, 8491.

3

Synthesis of Unsaturated Carbonyl Compounds

3.1. Aldol condensation/ Crossed aldol condensation

The union of two molecules of aldehyde **1** containing α H under acidic or basic condition to give β-hydroxyaldehyde **2** is called as aldol condensation. During the reaction small molecules such as water, alcohol or ammonia are eliminated.

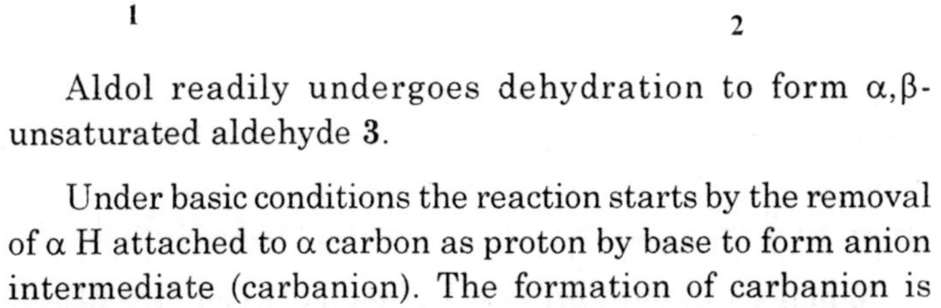

Aldol readily undergoes dehydration to form α,β-unsaturated aldehyde **3**.

Under basic conditions the reaction starts by the removal of α H attached to α carbon as proton by base to form anion intermediate (carbanion). The formation of carbanion is feasible because of presence of carbonyl group which helps in dispersing the –ve charge on carbon through resonance.

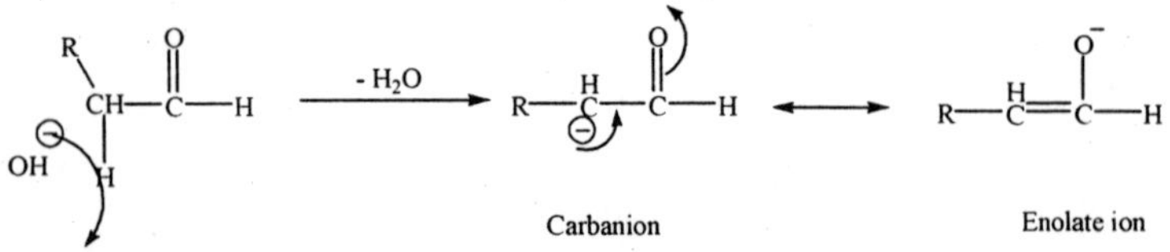

The anion is electron rich and behaves as nucleophile. The carbanion attacks the carbonyl carbon of another molecule of aldehyde to form **4** which takes a proton from solvent to give aldol **2**.

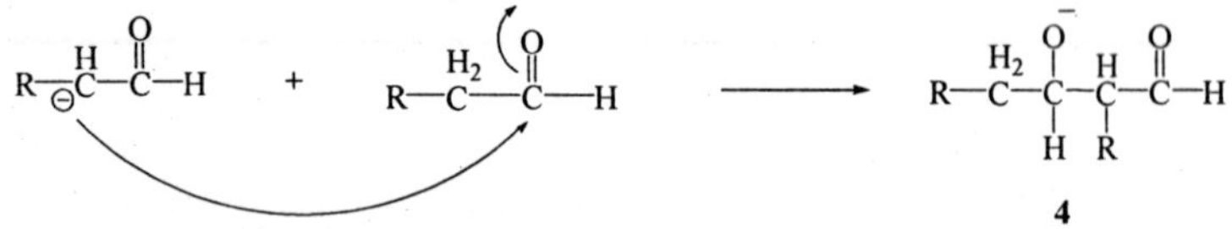

4

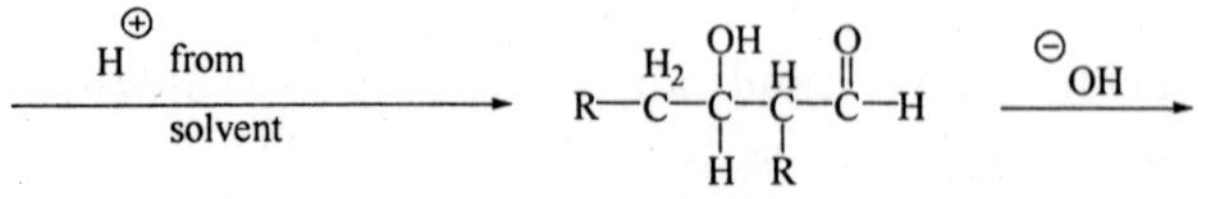

2

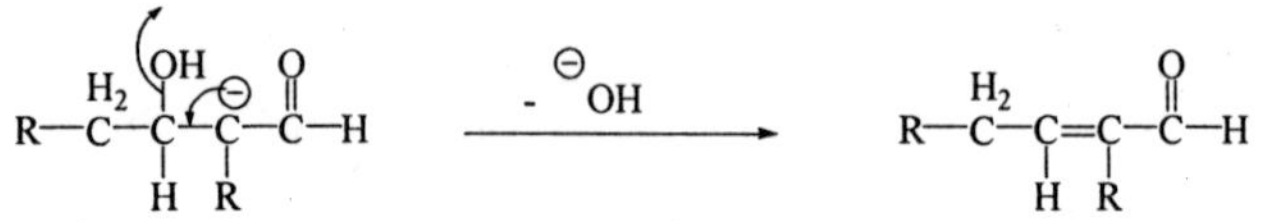

3

Aldol product is prone to dehydration. Thus, in the presence of base, a second deprotonation of the α carbon takes place and carbanion formed, helps in removing OH group to form an alkene, (α,β unsaturated aldehyde) **3**.

When condensation takes place between two different aldehydes or ketones, it is called as mixed or crossed aldol condensation.

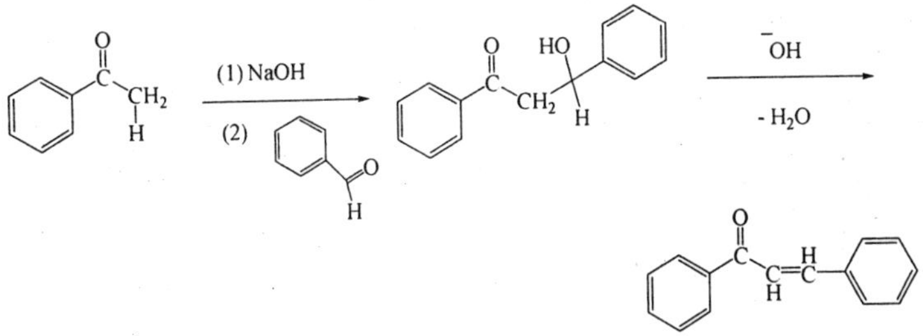

Different kinds of bases used in aldol condensation reaction are LDA, CH_3MgBr, Bu_2BOTf/NR_3, 9-BBN/NR_3, LDA + $ZnCl_2$, LDA + $(C_5H_5)_2ZrCl_2$, LDA + Ti(o-$^iPr)_3Cl$ whereas acids used are $TiCl_4$, $SnCl_4$, $AlCl_3$, BF_3. $O(C_2H_5)$, $ZnCl_2$ etc..[1,2]

3.2. Claisen Schmidt condensation

Condensation of an aromatic aldehyde **5** with an aldehyde or ketone having α-methyl group (aliphatic or aromatic) **6** in the presence of a strong base (sodium or potassium hydroxide) to form α, β-unsaturated aldehyde or ketone (chalcones) **7** is called as Claisen-Schmidt condensation.[3]

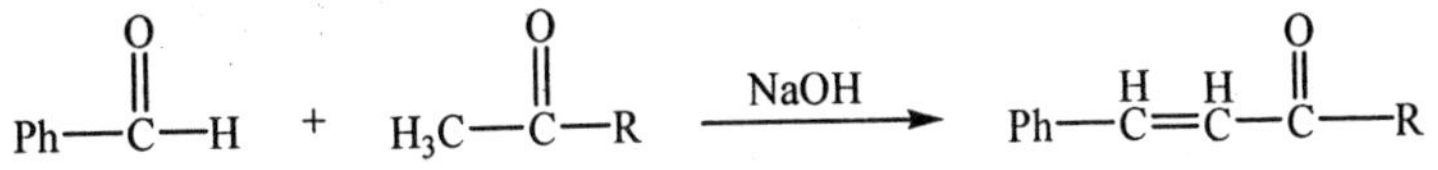

5 **6** **7**

Chalcone **7** acts as important intermediate for the synthesis of heterocycles such as pyrazoles, pyrazolines, and oxazoles etc.[4-6]

The mechanism involves formation of carbanion **8** from aldehyde or ketone containing α hydrogens.

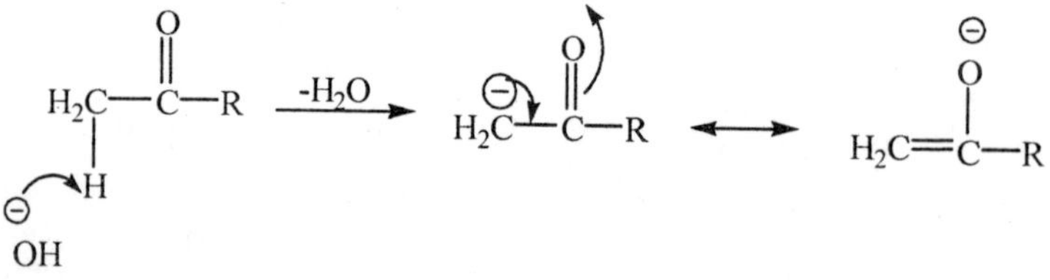

6 8

The carbanion **8** attacks carbonyl carbon of aromatic aldehyde to form 3-hydroxy compound **9** which subsequently undergoes dehydration to form the product **7**.

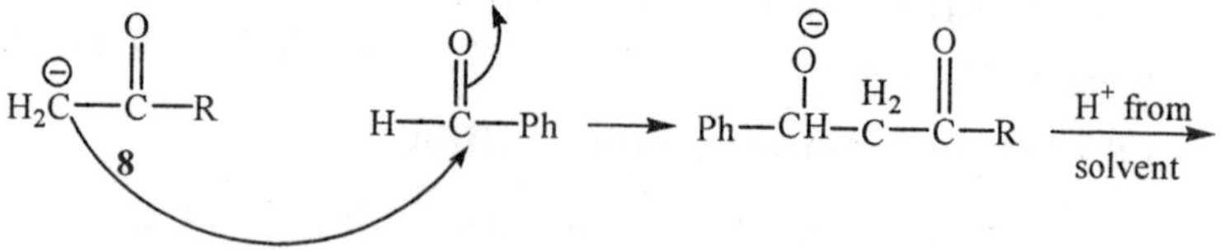

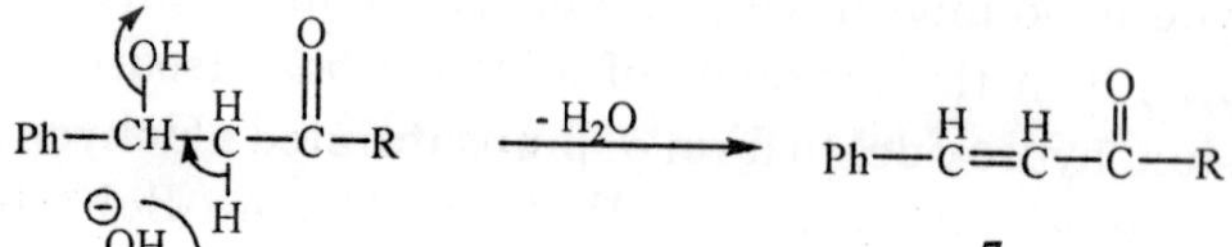

9 7

3.3. Robinson's Annulations

The reaction involves formation of a six membered ring containing α, β unsaturated ketonic functional group from an α, β unsaturated ketone derivative and a cyclic ketone containing α-H in the presence of a base. Thus, cyclic ketone

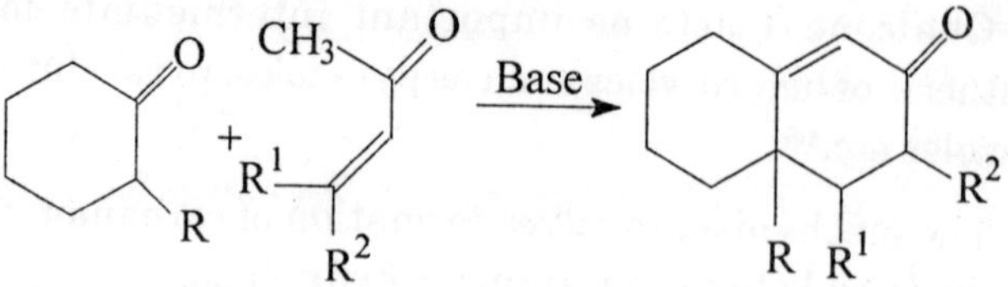

$R^1 = \quad R^2 = CH_3$
$R^1 = R^2 = H$
$R1 = H, = R^2 = CH_3$

12 has been obtained by Robinson's annulations employing cyclohexanone **10** and mehtyl vinyl ketone **11** in the presence of a base. The reaction involves formation of carbanion **13** of cyclohexanone with a base.[7,8]

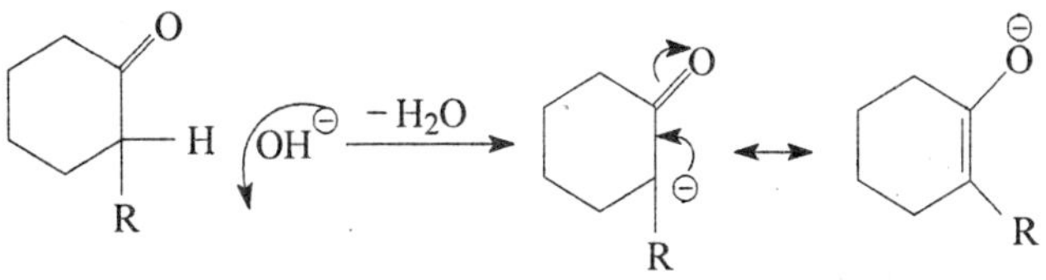

10 **13**

The carbanion **13** undergoes Michael addition with methyl vinyl ketone to give

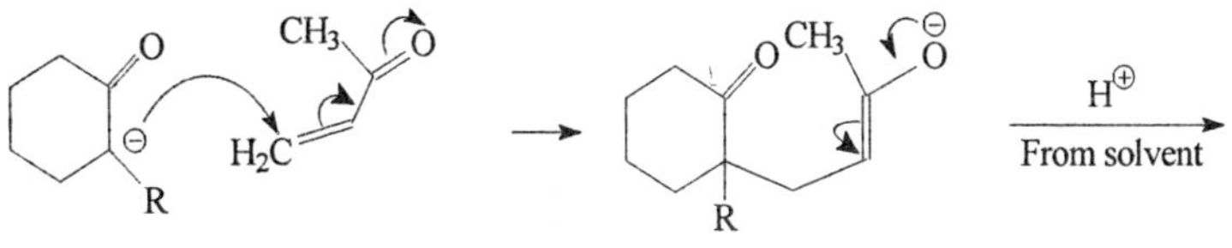

13 **11** **14**

condensed product **14** which undergoes further aldol condensation followed by dehydration to give **12.**

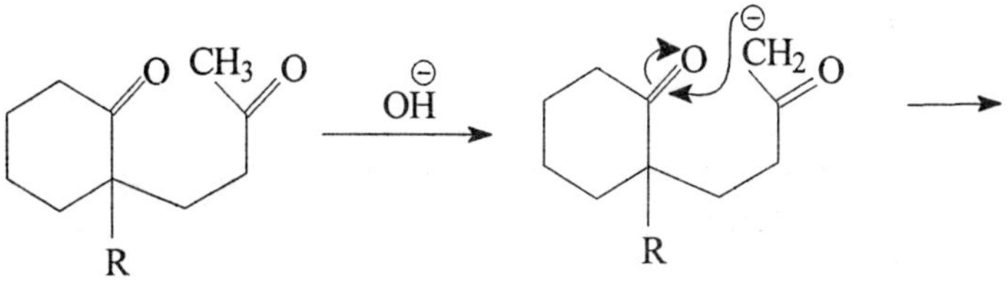

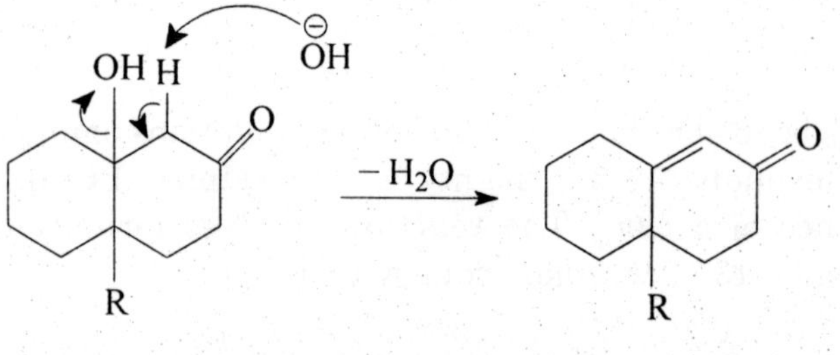

12

Thus, cyclohexanone derivative **17** has been synthesized from ketone **16** and α, β unsaturated ketone **15** in the presence of base La [OPr^{i})$_3$.[9]

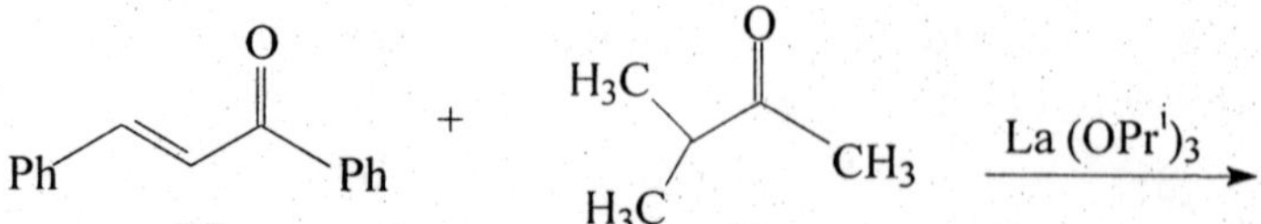

15 **16**

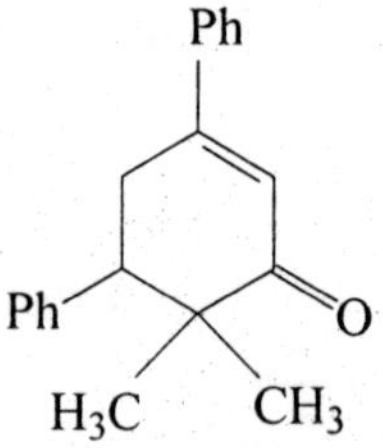

17

The Robinson's annulation catalyzed by antibody has been also reported.[10]

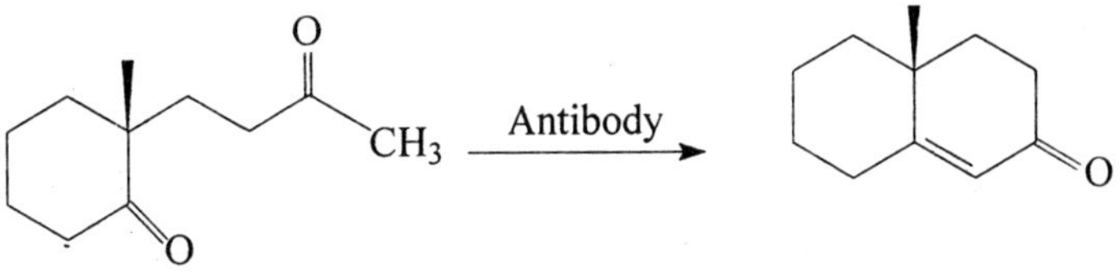

REFERENCES

1. T. Mukaiyama, *Org. React.* ,1982, **28**, 203.
2. C.H. Heathcock, *Comp. Org. Syn.*, 1991, **2**, 181.
3. L. Claisen and A. Claparede, *Ber.*, 1881 ,**14**, 2460, J.G. Schmidt, *ibid*, 1459.
4. M.G. Joshi and K. N. Wadodkar, *Indian J. Chem.* 1981, **20B**, 1090.
5. A.A. H. Abdel-Rahman, A. E. S. Abdel-Megied, M. A.M. Hawata, E. R. Kasem and M. T. Shabaan, *Monatsh. Chem.*, 2007, **138**, 889.
6. V.D. Orlov, N.P. Vorobeva, N. N. Dernenkova, V.S. Chesnovki and F.G. Yaramanko, *Khirn. Geterotsikl. Soedin.* 1988, **3**, 328.
7. E.C. du Feu, F.J. McQuillin and R. Robinson, *J. Chem. Soc.*, 1937, 53.
8. K. Balasubramanian, J.P. John and S. Swaminathan, *Synthesis*, 1974, 51.
9. T. Okano, Y. Satou, M. Tamura and J. Kiji, *Bull. Chem.Soc. Jpn.* ,1997, **70**, 1879.
10. G. Zhong, T. Hoffmann, R.A. Lerner, S. Danishefsky and C.F. Barbas III, *J. Am. Chem. Soc.* ,1997, 119, 8131.

4

Synthesis of Flavonoids

4.1. Algar-Flynn-Oyamada reaction

Oxidation of o-hydroxy chalcones **1** with alkaline hydrogen peroxide forms flavonol **3** via the intermediate dihydro flavonol **2**.[1,2]

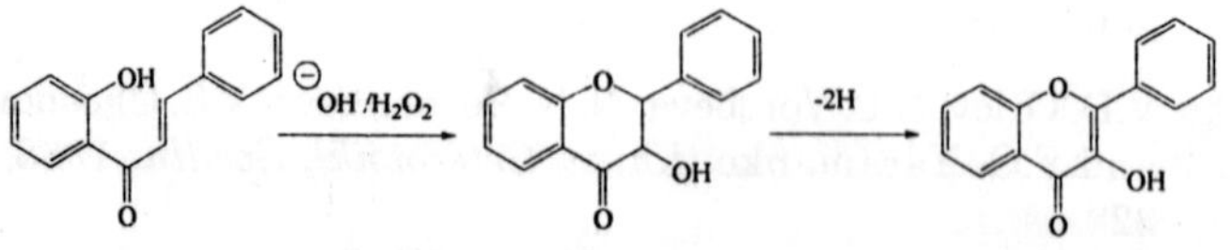

1 **2** **3**

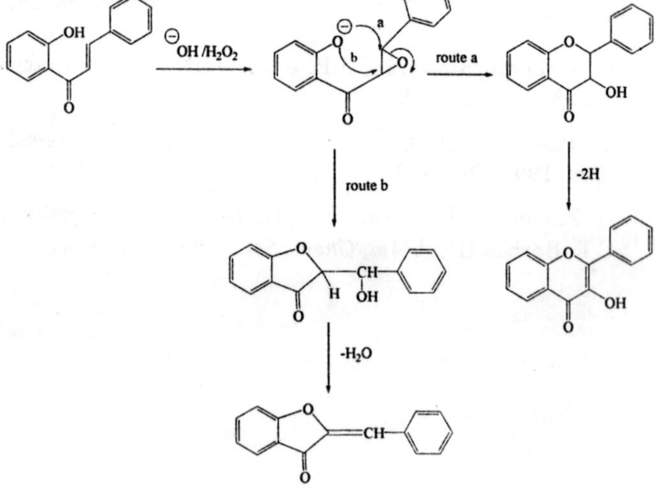

O. Sullivan et.al[3] have reported that 2'- tosyloxy chalcone epoxide **4** reacts with alkali to give flavonol **5**. However, when 6'-methoxy-2'-tosyloxy chalcone **6** is heated with alkali, both at room temperature and reflux temperature ennone **7** rather than flavonol **5** is obtained.

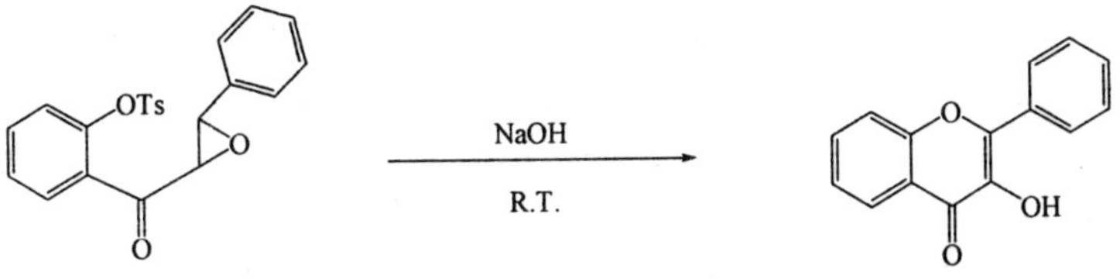

4 **5**

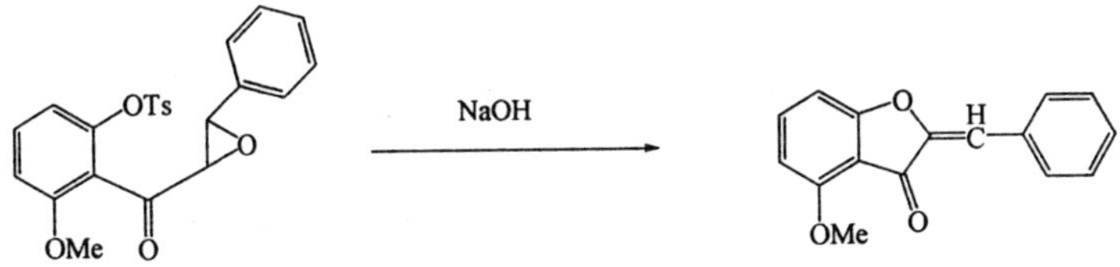

7

6

One of the reasons suggested by the authors is the displacement of the carbonyl group out of the plane of the phenolic ring by steric interaction with the methoxyl group at 6' position increases the distance of the phenolic oxygen from β position more than from α position.

2,2-Dimethyl-2H-pyran fused flavonols **11** have been synthesized by using modified Algar-Flynn-Oyamada reaction.[4]

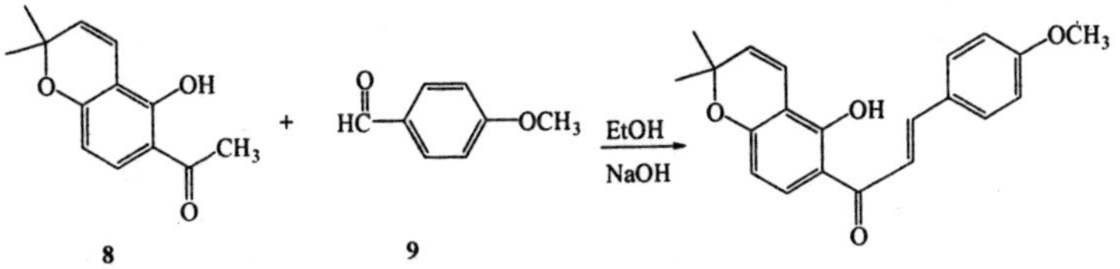

8 **9**

10

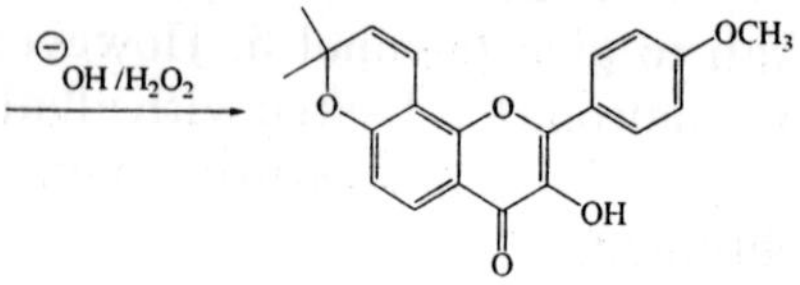

11

The formation of pyran fused flavonol **11** involves treatment of 6-acetyl-2,2-dimethylchromenes **8** with p-anisaldehyde **9** in the presence of ethanolic NaOH and subsequent oxidation with alkaline hydrogen peroxide without isolating the intermediate chalcone **10**.

4.2. Allan- Robinsons reaction

Allan-Robinson reaction is used for the synthesis of flavones and isoflavones by the reaction of o-hydroxyarylketones with anhydrides of aromatic acids in the presence of sodium salt.

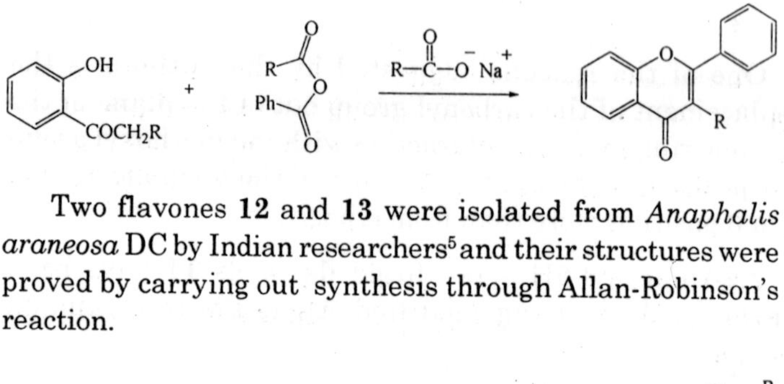

Two flavones **12** and **13** were isolated from *Anaphalis araneosa* DC by Indian researchers[5] and their structures were proved by carrying out synthesis through Allan-Robinson's reaction.

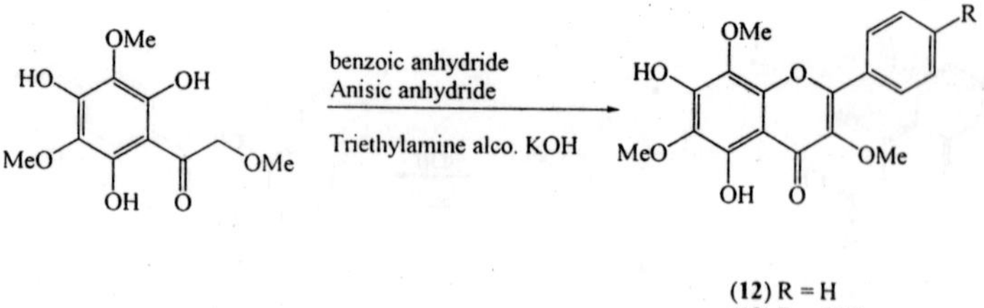

(12) R = H
(13) R = OMe

The flavonol **15** isolated from citrus limon (lemon peel) was synthesized by Dreyer et. al. through Allan- Robinson reaction from **14**.[6]

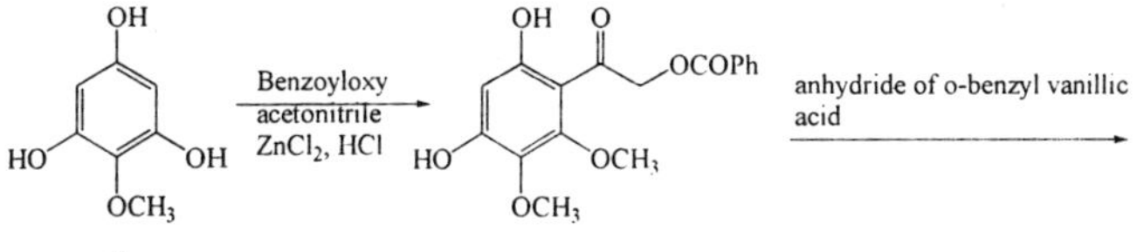

14

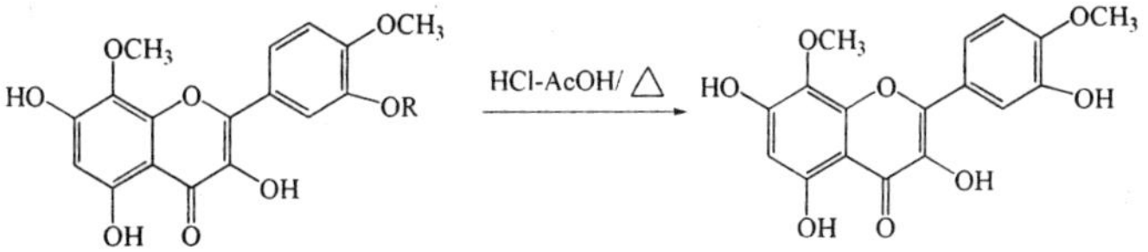

15

4.3. Auwers Synthesis

The reaction involves synthesis of flavones from *o*-hydroxy chalcones or coumaranone (benzofuran-3-ones). Synthesis from benzofuran-3-one **16** involves ring expansion in the presence of a base.[7,8]

The reaction involve following steps

(i) Formation of 2-benzylidene **17** by the reaction of coumaranone **16** and benzaldehyde under acidic conditions.

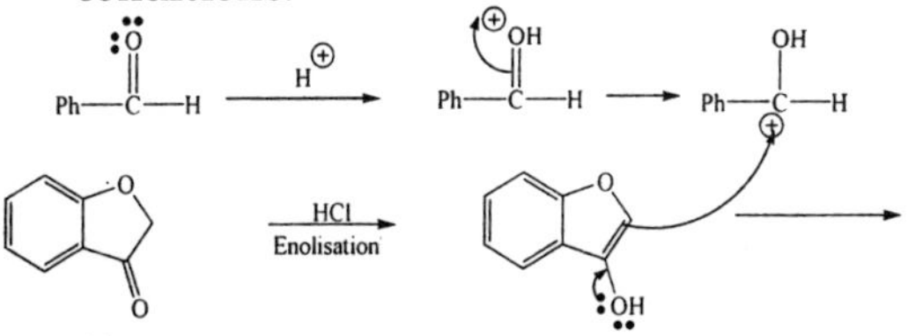

16

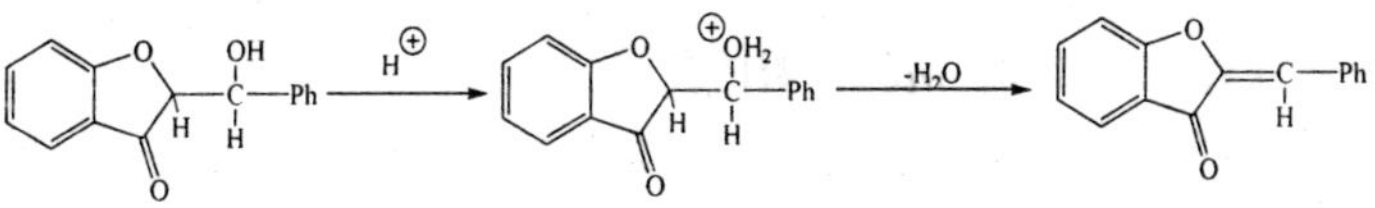

17

(ii) Bromination of benzylidine **17** to give dibromocompound **18**.

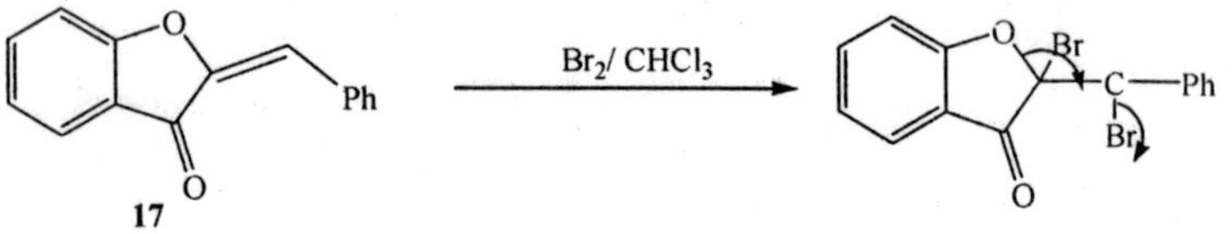

(iii) Ring expansion of dibromocompound **18** upon treatment with alcoholic KOH takes place to give flavonol **19**.

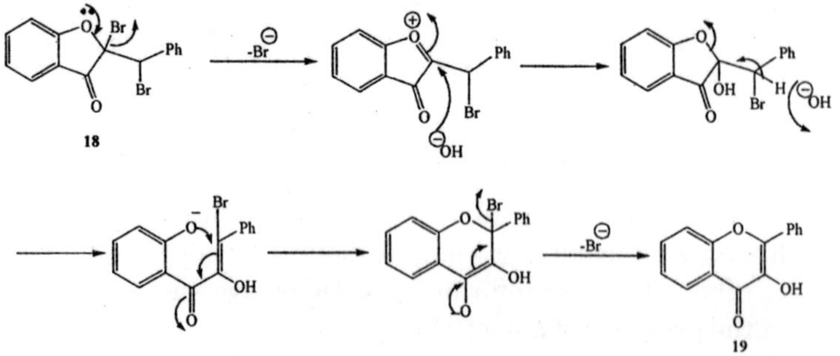

REFERENCES

1. M. K. Seikel, *J. Am. Chem. Soc.*,1955, **77**, 5685.
2. T. A. Geissman and J. B. Harborne, *Arch. Biochem. Biophys.*,1955, **55**, 447.
3. T. R. Gormley and W. I. O' Sullivan, *Tetrahedron,* 1973, 29, 369.
4. A.C.Jain, S.M.Gupta and A. Sharma, *Bull. Chem. Soc. Jpn.*, 1983, **56**,1267.
5. P. K. Dutta, D. Bagchi and S. C. Pakrashi, *Indian J. Chem.*,1982, **21B**, 1037.
6. D. L. Dreyer, S. Tabata and R.M. Horowitz, *Tetrahedron,* 1964, **20**, 2977.
7. K. Auwer, *Chem. Ber.*,1908, **41**, 4233.
8. K. Auwers and P. Pohl, *Ann.* ,1914, **405,** 243.

5

Synthesis of Amines

5.1. Amadori Rearrangement

This rearrangement has been observed in sugars.[1] It involves isomerisation of a glycosylamine to an aminosugar. The reaction takes place in the presence of a base at pH 11. The isomerisation occurs at room temperature and proceeds through enolic intermediate **2**.

Thus, D-fructosylamine **1** isomerizes to D-glucosamine **3** in liquid ammonia containing NH_4OH.

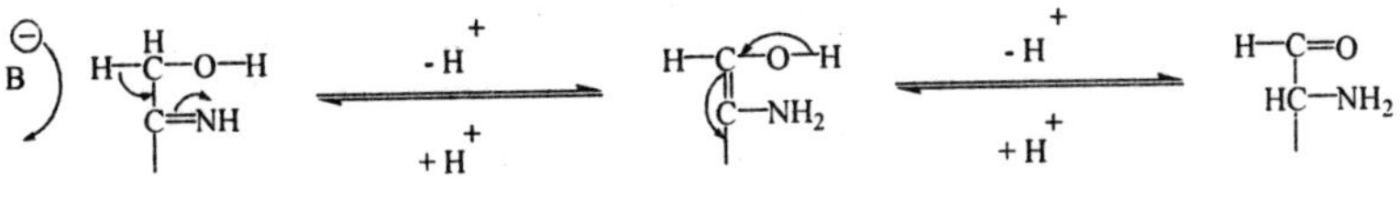

1 2 3

The authors, Isbell and Frush[2] have explained the mechanism which is catalysed by both acids and bases. The reaction is also strongly catalysed by active methylene compounds[3] such as diethyl malonate, acetyl acetone and carboxylic acids.[4]

The authors proposed that the condensation of the compounds with an intermediate imminium ion from the glycosylamine takes place. Thus, opening of the ring of the glycosyl amine results, which then enolizes to enolamine.

Glycosylamine salts must be easily hydrolysed in aquous solution. However, nonaquous conditions involving alcohol containing an acid are usually used for Amadori rearrangement.

5.2. Bischler Mohlau indole synthesis

The reaction involves synthesis of indole **6** from α bromoacetophenone **5** and aniline **4**.

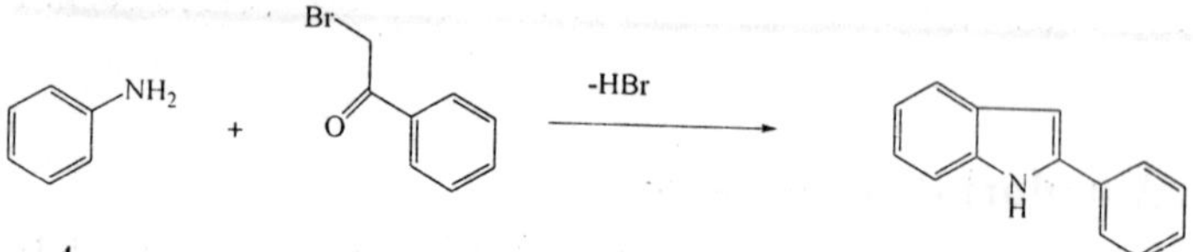

4 5 6

The reaction is initiated by the nucleophilic attack of NH_2 group of aniline **4** on α carbon bearing bromine atom to form **7** which undergoes cyclization to form the product **6**.

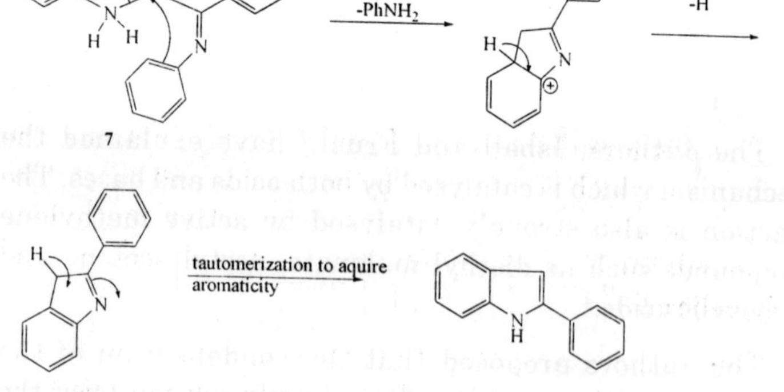

7

6

5.3. Bischler Napieralski reaction

It is a chemical reaction used for the synthesis of isoquinoline. The reaction takes place by dehydration of N-acyl β-aryl ethyl amine **8** to isoquinoline **9** in the presence of dehydrating agents such as P_2O_5, PCl_5, poly phosphoric acid (PPA), $ZnCl_2$, $POCl_3$ etc.[5]

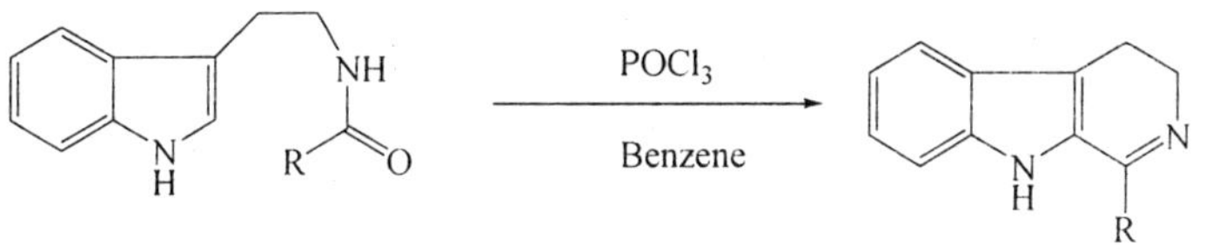

8 **9**

Some important starting molecules leading to the formation of isoquinoline are shown below.

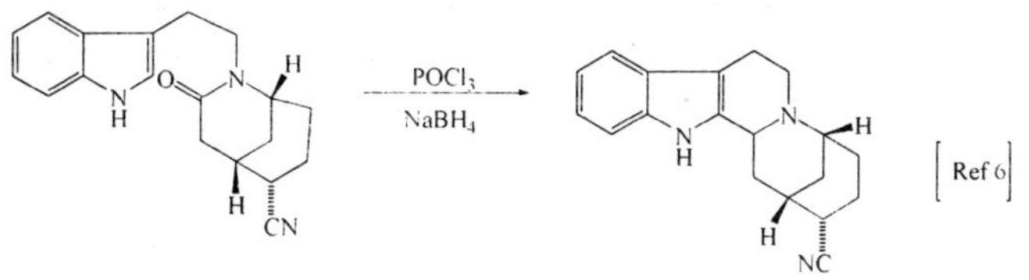

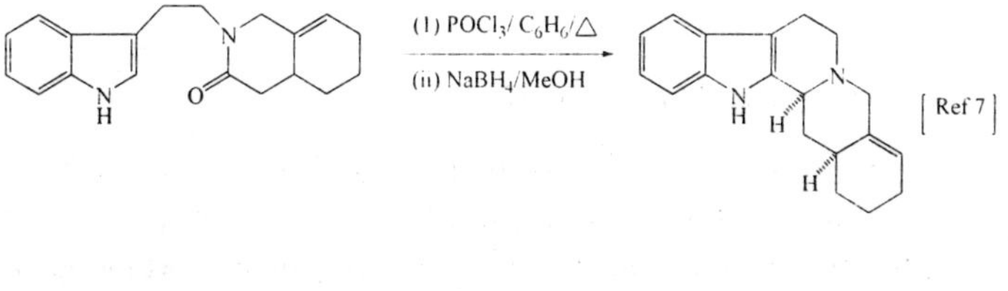

The mechanism of the reaction involves formation of intermediate **10** which undergoes cyclization to form the product **11**.

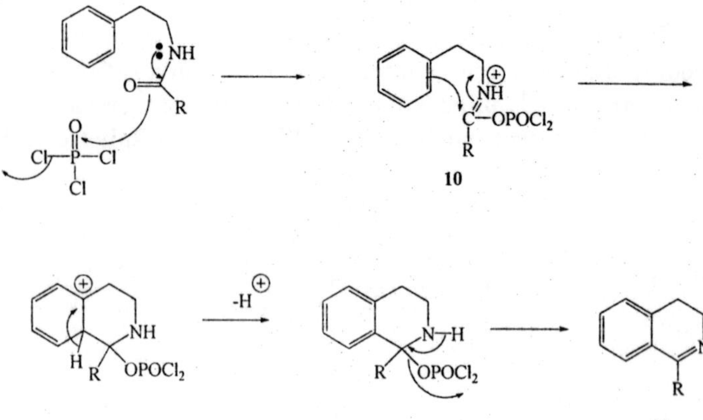

10

11

5.4. Chichibabin pyridine synthesis

The reaction is used for the synthesis of pyridine derivatives. The reaction consists in condensing carbonyl compounds (aldehydes or ketones) with ammonia and amines under pressure.[8] Thus, aldehyde **12** reacts with NH_3 to form **13**.

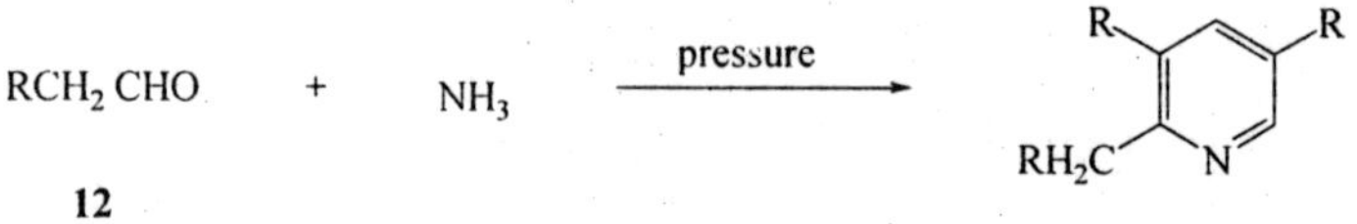

12

13

5.5. Chichibabin reaction

Introduction of NH_2 group to N-heterocycles such as pyridine, quinoline by means of alkali metal amide is called as Chichibabin reaction. Usually the amination takes place at α-position (C-2).[9] If both α-positions (C-2 and C-6) are having substituents then NH_2 group enters on γ position (C-4). Thus, when pyridine **14** is heated with sodamide in toluene at 100 ° C, 2-amino pyridine **15** is obtained.

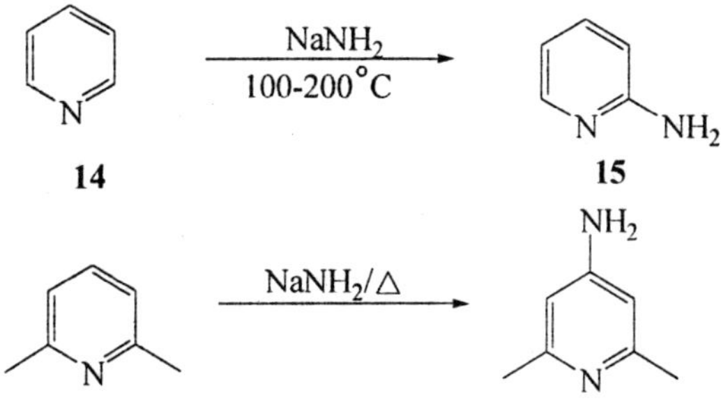

The mechanism of the reaction involves nucleophilic substitution reaction. NH_2 group preferably attacks at α position to form **16** which removes a hydride ion to give**17**.

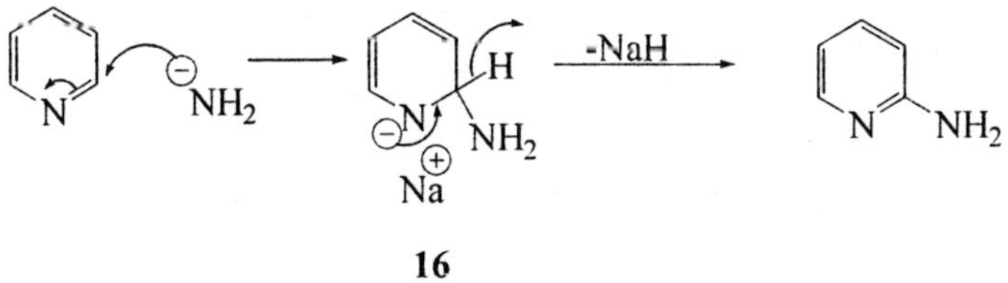

The final product, 2-amino pyridine **15** is obtained after hydrolysis of salt **17**.

Chichibabin type reaction using amino acid

3,5-Diaryl pyridine **19** has been synthesized by modified chichibabin reaction. The product 3,5-diarylpyridine **19** is obtained by the reaction of 3-nitrotyrosine **18** with hypochlorous acid (HOCl) in aquous phosphate buffer.[10]

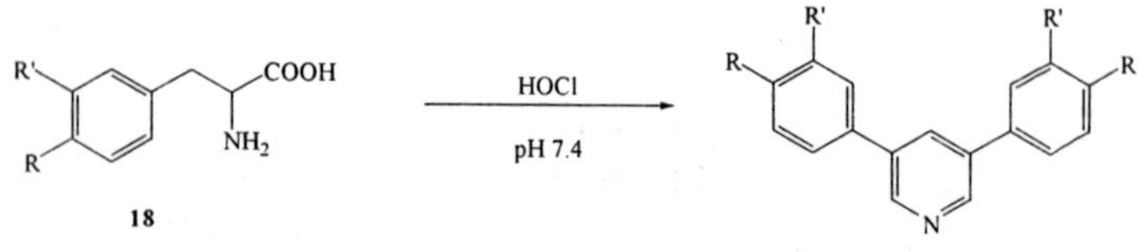

1. R, R' = H
2. R= OH, R' = H
3. R= OH, R'= NO_2

19

5.6. Combes quinoline synthesis

The reaction is used for the synthesis of quinolines. The reaction takes place when β-diketone is condensed with primary aryl amine to form the intermediate **20** which undergoes cyclization in the presence of sulfuric acid to afford the product **21**.[11]

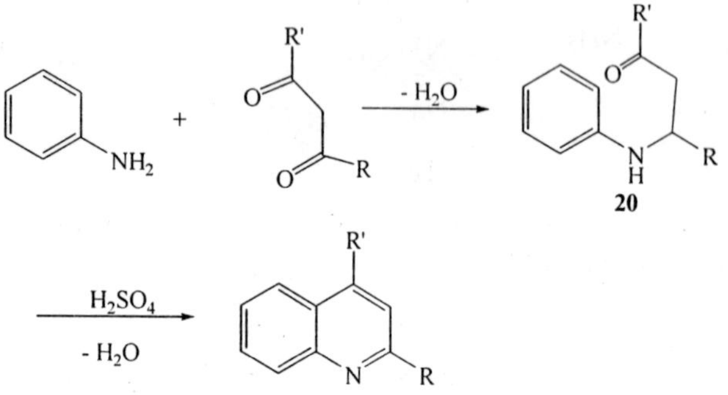

21

The mechanism involves following steps.

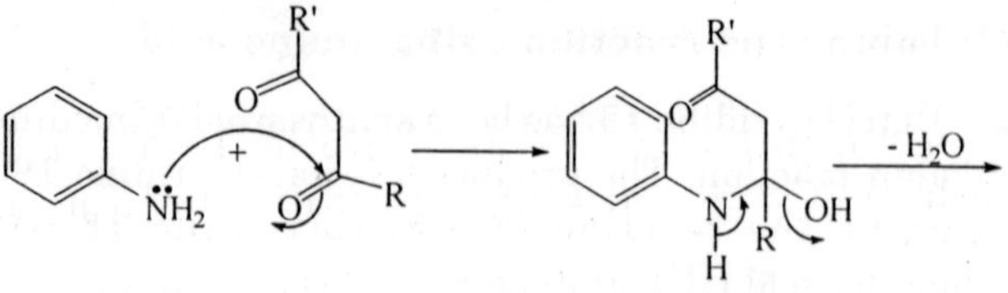

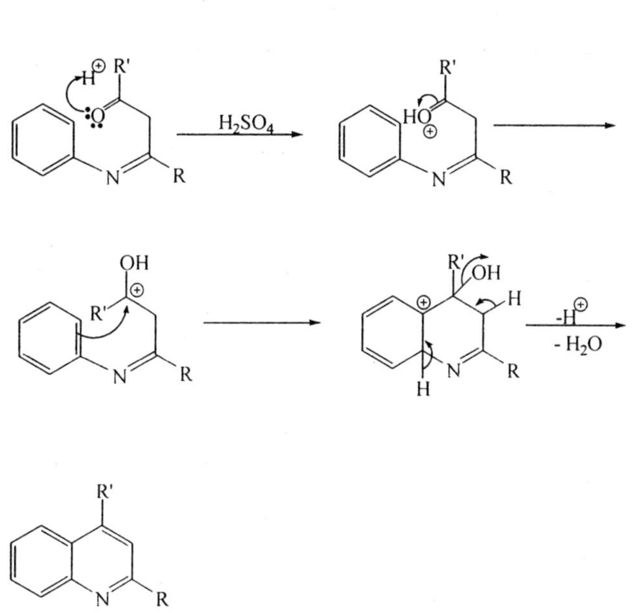

5.7. Doebner Miller reaction

The reaction involves treatment of α,β-unsaturated carbonyl compound **23** with primary aromatic amine **22** to form quinolines **24**. The reaction takes place in the presence of some acid catalyst.[12]

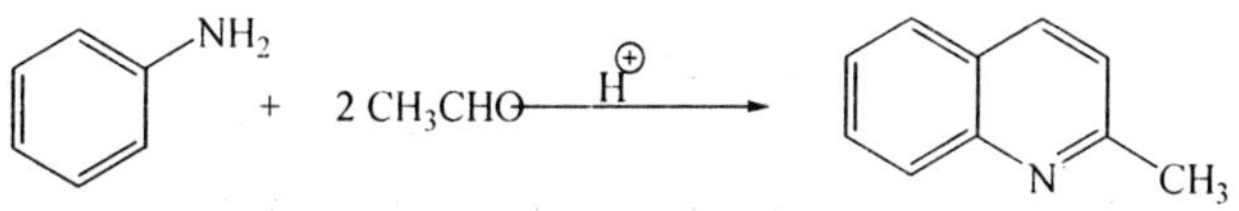

22 **23** **24**

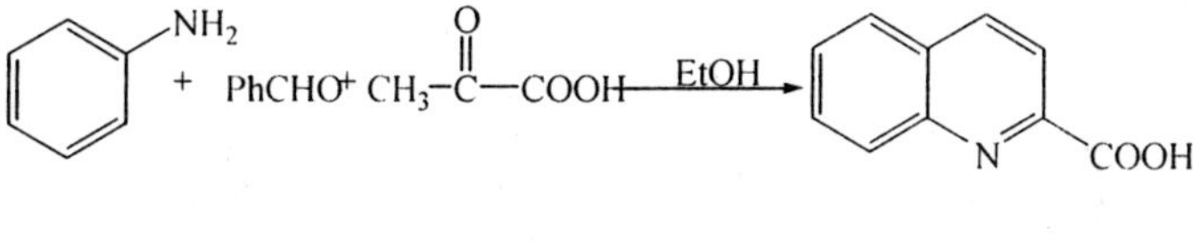

The plausible mechanism is outline below.

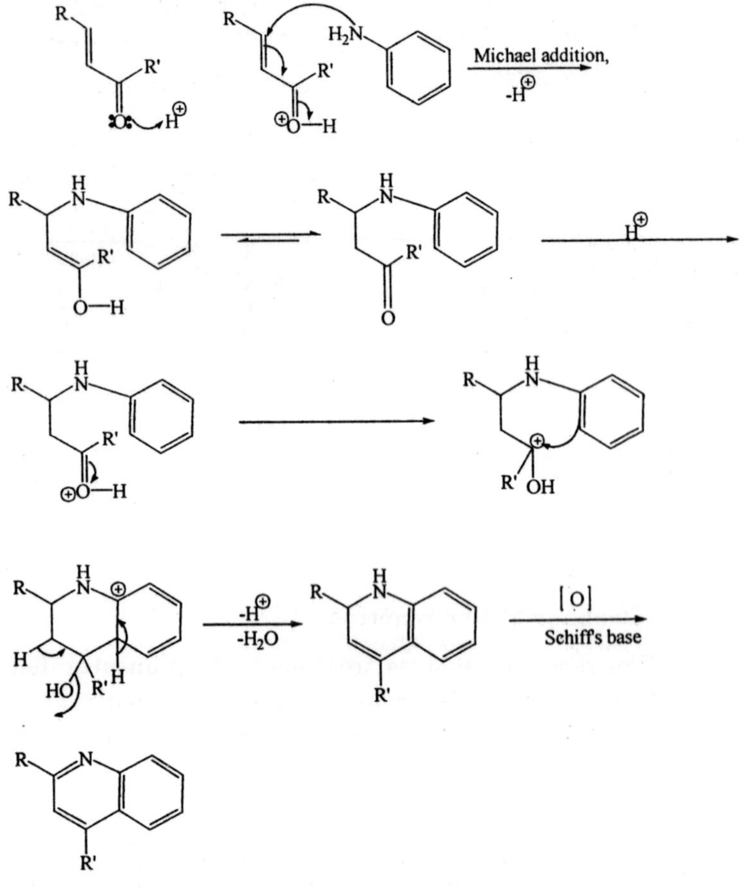

The α, β-unsaturated carbonyl compound may be obtained by aldol condensation of carbonyl compounds containing α-methyl ketone groups.

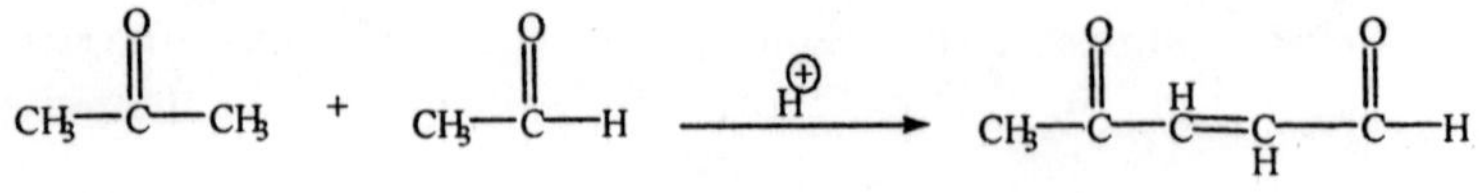

It is believed that oxidation of dihydroquinoline is brought out by schiff's base which is produced *in situ* by the reaction of amine and carbonyl compound.

Synthetic application of Doebner Miller reaction

(i) During Doebner Miller reaction, 2,6- dimethyl-1,2-dihydroquinoline has been isolated. Further, dihydroquinoline **26** is prepared by acid catalysed dehydration of 4-hydroxy-1,2,3,4-tetrahydroquinoline **25**.

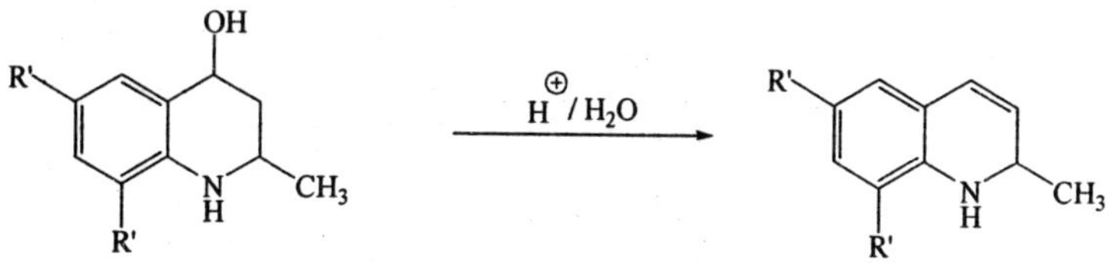

25 **26**

(ii) *Synthesis of quinalidines and lepidines by Doebner Miller reaction under thermal as well as microwave irradiation*: Quinalidines and lepidines have been synthesized by one pot reaction of anilines with crotonaldehyde or methylvinyl ketone using phosphotungstic acid (a Keggins type hetero polyacid) under both thermal and microwave irradiation conditions.[14]

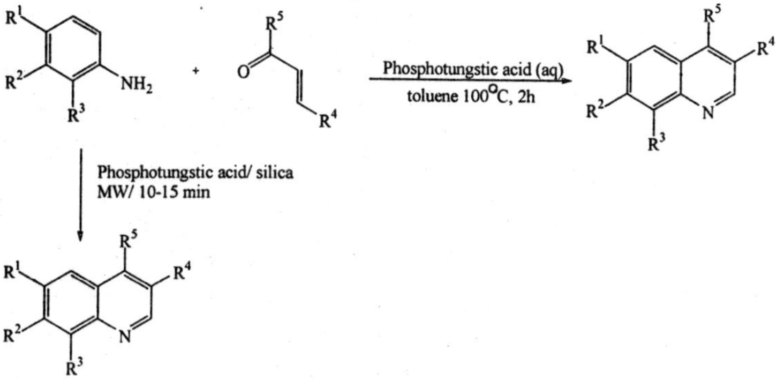

5.8. Fischer indole synthesis

The reaction used for the synthesis of indole **27** by heating hydrazone of an aldehyde or ketone in the presence of a Lewis acid is called as Fischer- Indole Synthesis.[15]

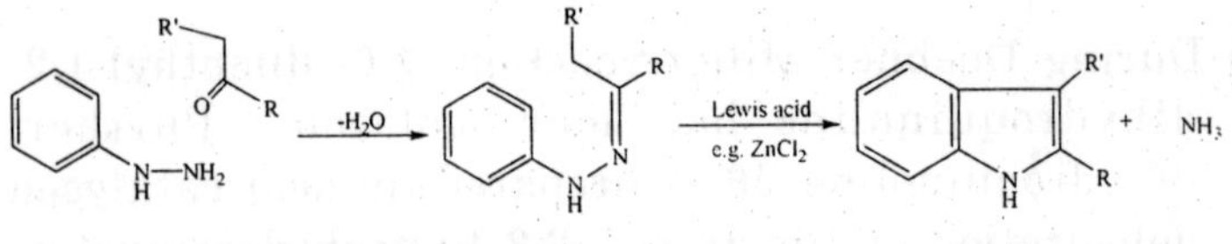

27

The conversion of arylhydrazone to indole usually requires high temperature.

In the formation of product, indole, disproportanation of proton takes place at various stages of the reaction.

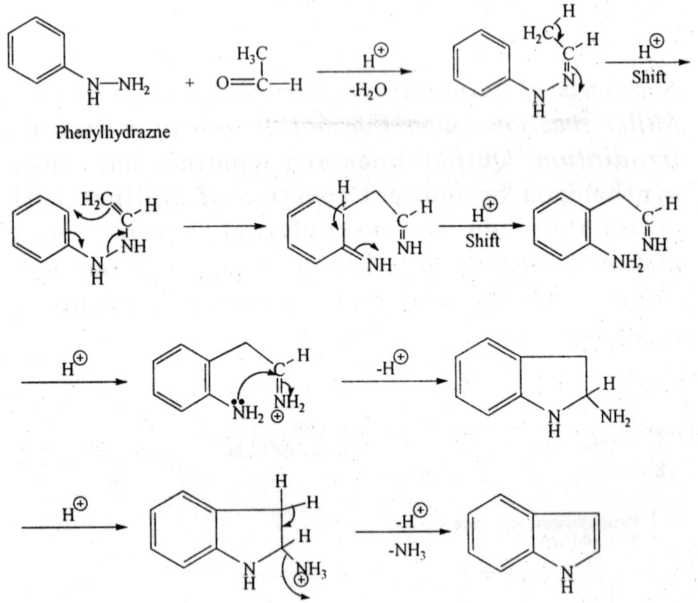

Some recent examples involving synthesis of indole

1. *Fischer indole synthesis has been carried out using palladium as catalyst.*[16]

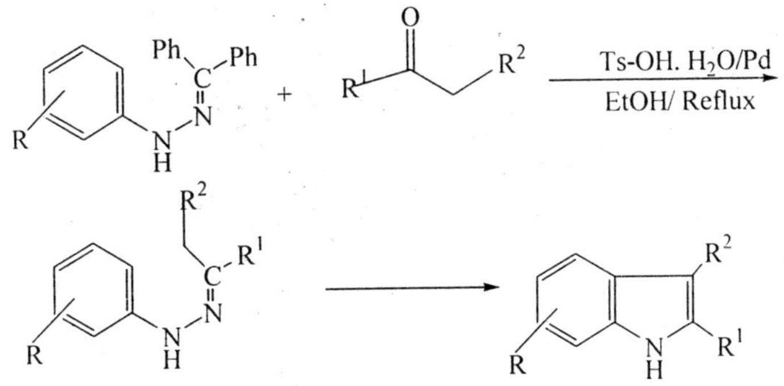

2. *Fischer Indole Synthesis in Bronsted acidic ionic liquid:*

A novel one pot Fischer Indole method for synthesizing indole was developed by Z. Y. Xu et.al.[17] using Bronsted acidic liquids as dual solvent catalyst and employing alkyl methyl unsymmetrical ketones as starting material. Yields were quantitative and the procedure was simple involving easy work of the products from the reaction mixture.

3. *Fisher indole synthesis on solid support*

2- Aryl indoles have been also synthesized on a solid support.[18]

The reaction involves support bound 4-benzoyl butyric acid and variously substituted phenyl hydrazine hydrochlorides as the synthetic building blocks.

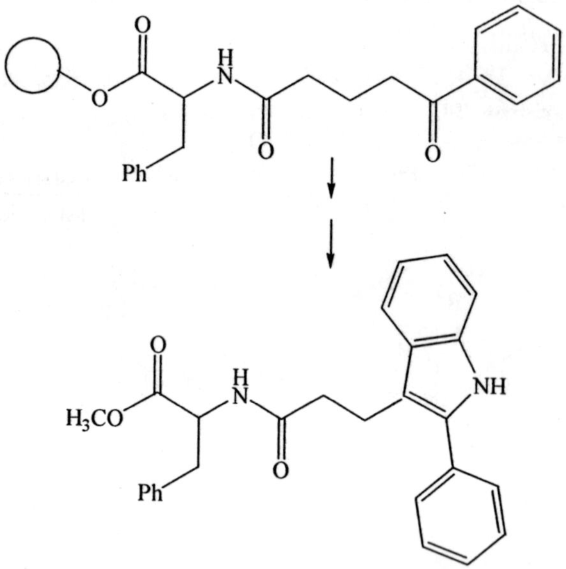

4. *Thermal cyclization of N-trifluroacetylenehydrazine under mild conditions forms indole.*[19]

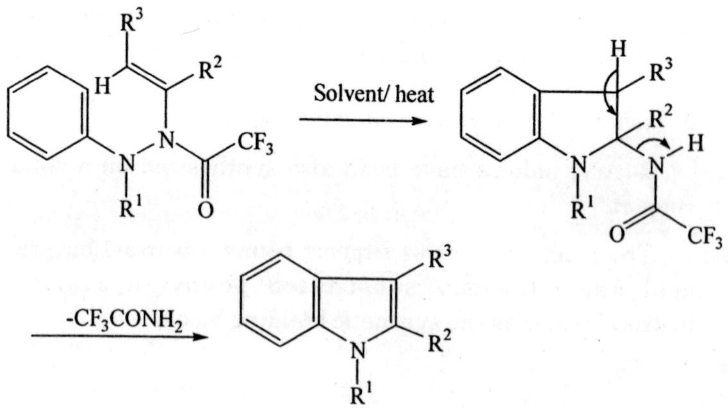

5.9. Hantzsch Dihydropyridine synthesis

The reaction is used for the synthesis of dihydropyridine derivative. Dehydrogenation of dihydropyridine derivative gives pyridine. The reaction consists in treating an aldehyde, active methylene compound in the presence of ammonia.[20]

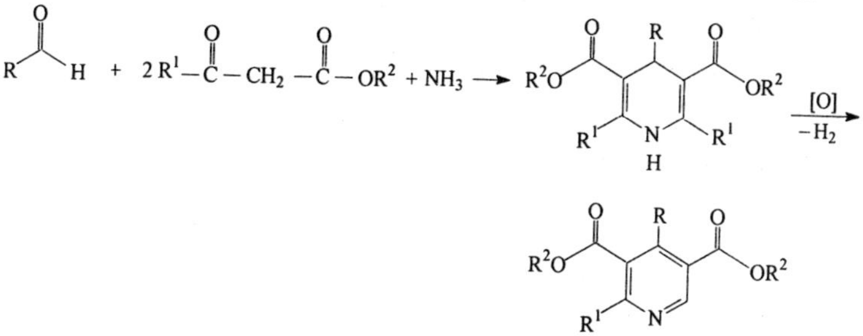

When aldehyde is replaced by chloroketone then pyrrole is obtained.

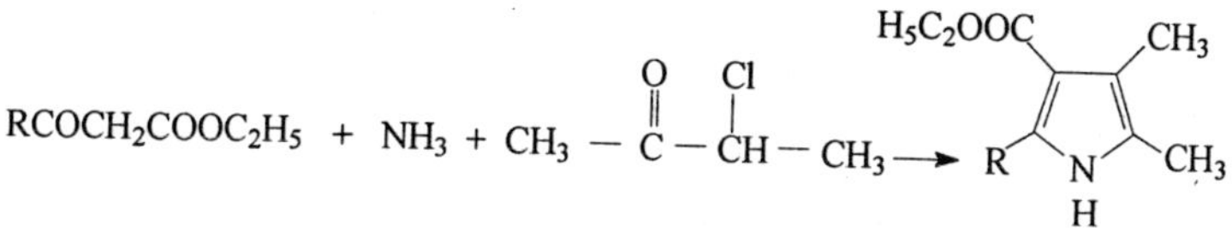

The mechanism of the reaction takes place as shown below.

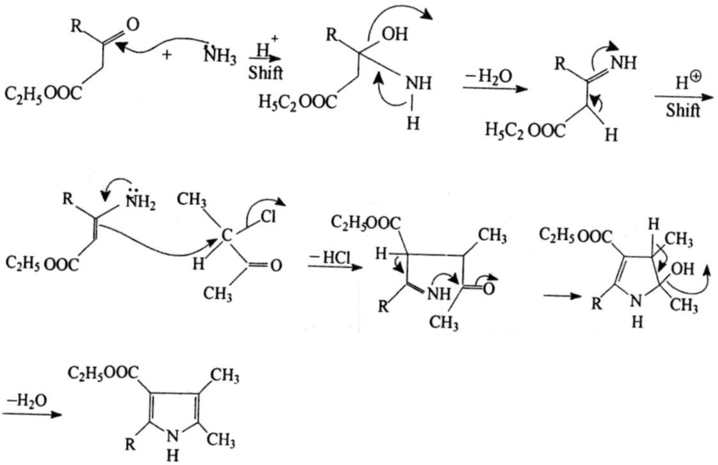

5.10. Friedlander Synthesis

Base catalysed condensation of o-amino benzaldehyde with enolizable ketones to form quinoline derivatives is known as Friedlander synthesis.[21]

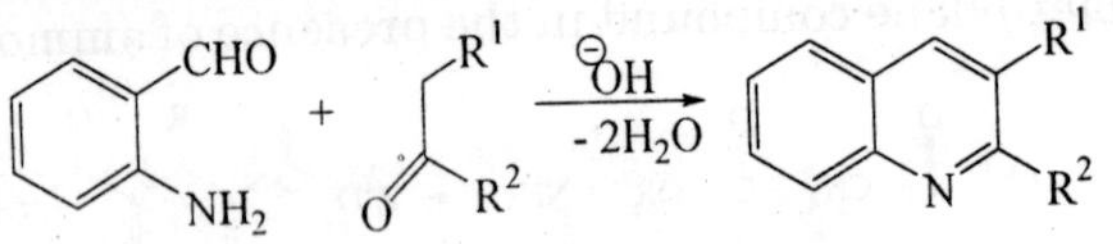

The mechanism of the reaction involves formation of carbanion by abstraction of acidic proton of aldehydes/ketone with a base. The carbanion **28** attacks aldehydic carbon of *o*-amino benzaldehyde to form **29** which undergoes cyclodehydration to give the product **30**.

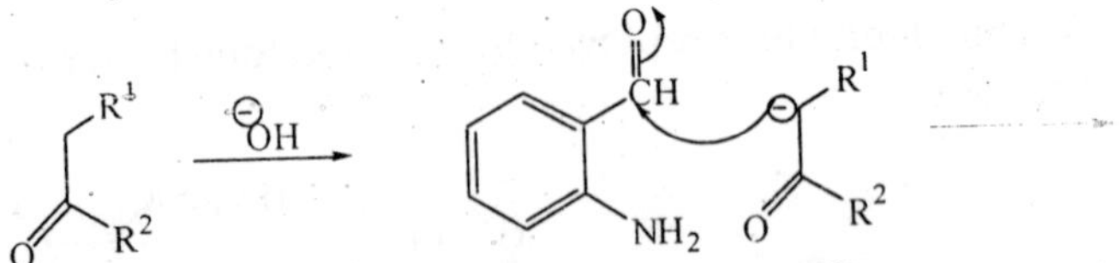

28

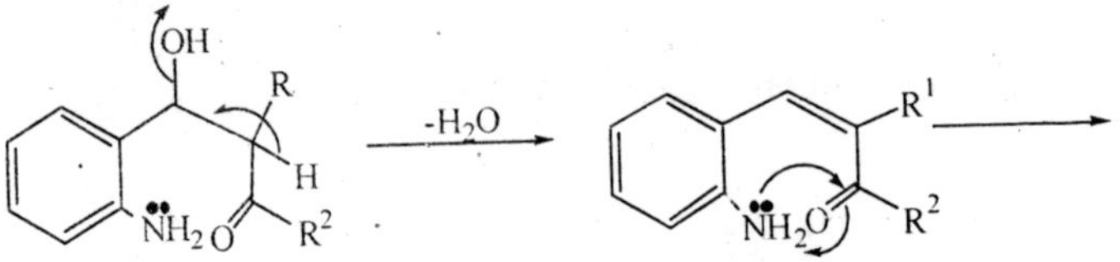

29

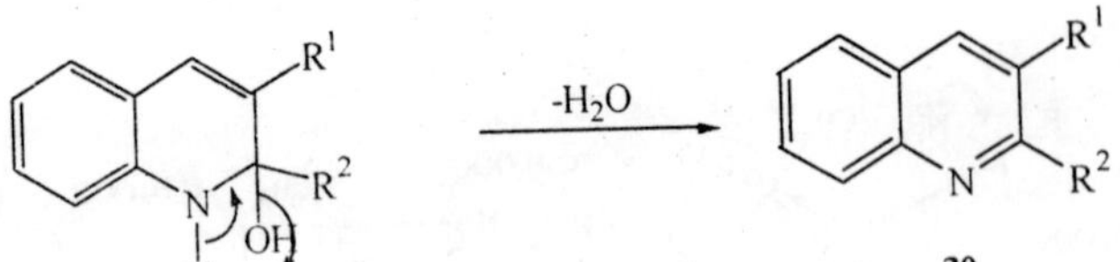

30

Synthetic application of Friedlander reaction

1. *Microwave assisted Friedlander condensation catalysed by clay.*

In recent years microwave assisted reactions are gaining attention of chemists. They are beneficial in many ways over conventional approach because of short reaction time, cleaner reaction, with easy work up. Friedlander condensation has been reported under microwave irradiation by carrying out reaction of o-aminoaryl aldehydes or ketones with carbonyl compounds containing α-methylene group catalysed by clay in solvent free conditions to give quinoline derivatives.[22]

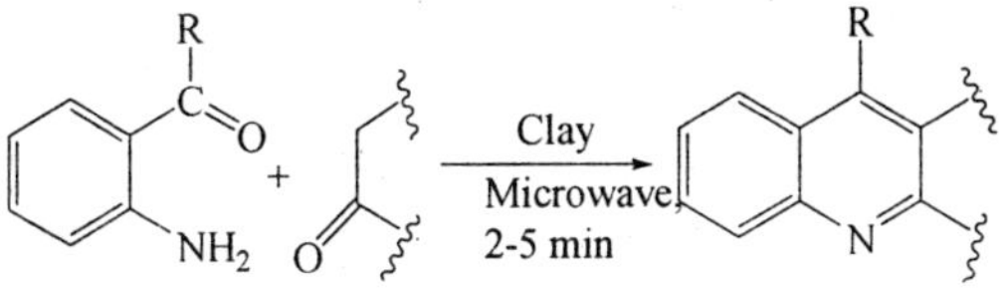

2. *Synthesis of 11 H- Indeno[1,2-b] quinolines via Friedlander condensations reactions*. Synthesis of quinoline derivatives through Friedlander condensation has been reported by carrying out reaction of *o*-amino benzaldehyde **31** with indenone **32**. The product **33** is obtained in quantitative yield. The reaction is usually catalysed by the use of sodium ethoxide as base.[23]

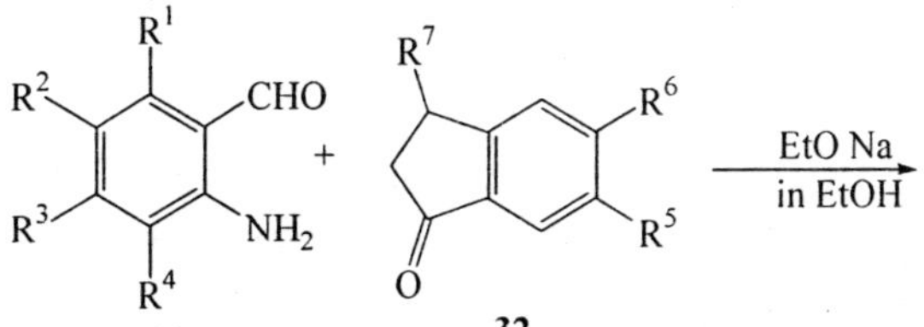

31 **32**

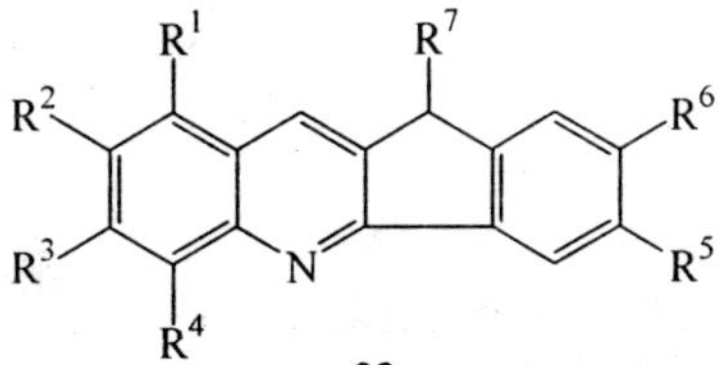

33

The mechanism of the reaction involves following steps:

(i) Formation of resonance stabilized carbanion **34** by removal of α proton from indenone **32**.

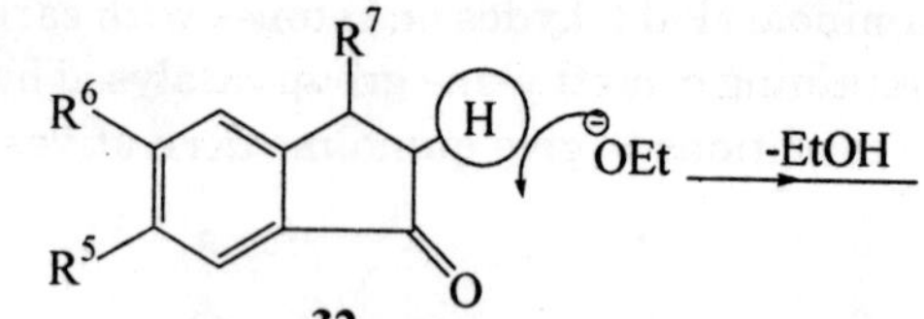

32

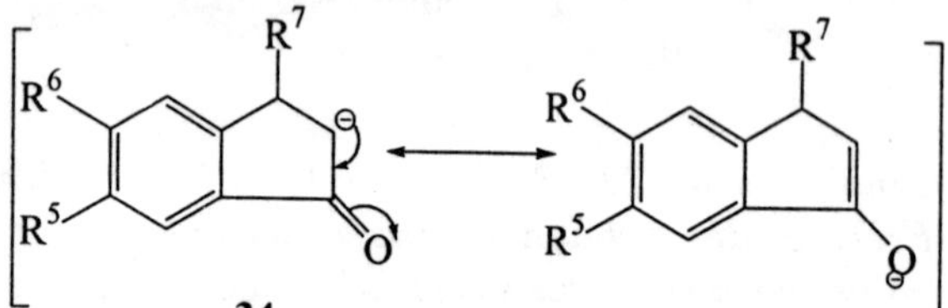

34

(ii) Attack of carbanion **34** on aldehydic carbon of **31** takes place to form **35**. Subsequent cyclisation by amino group takes place which is followed by dehydration to afford the product **33**.

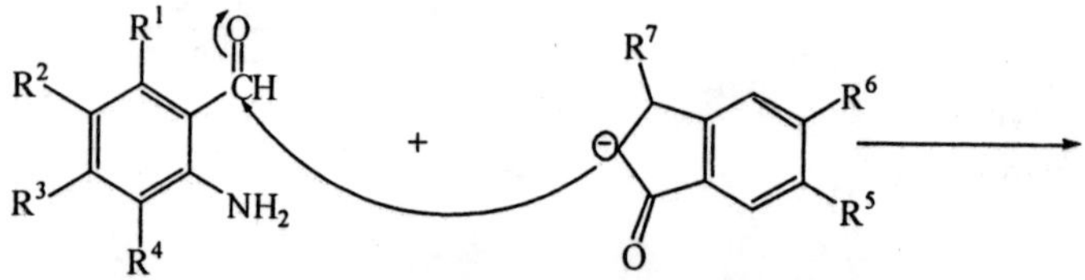

34

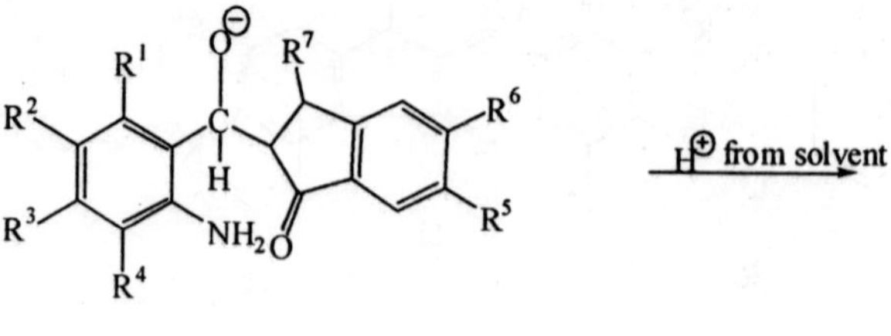

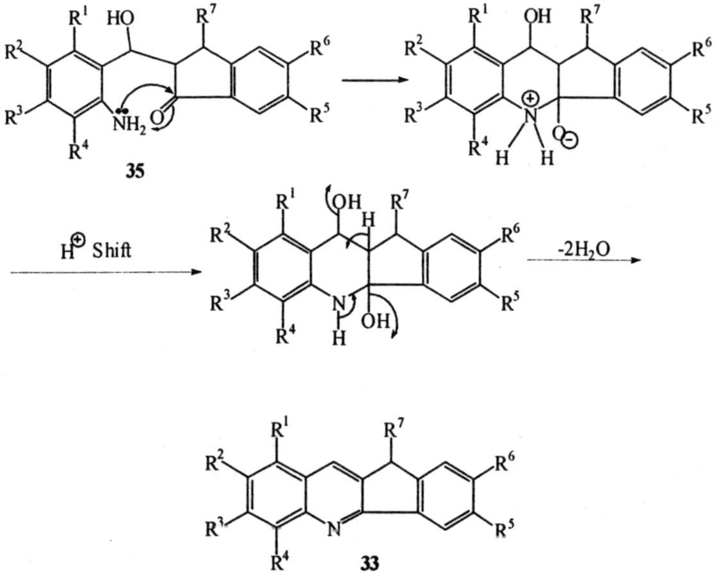

3. *Synthesis of 4-aryl-1H-pyrazolo[3,4-b] quinoline by cyclization of 4-arylidene pyrazolin5-ones with anilines.*

This reaction which does not involves Friedlander condensation describes synthesis of pyrazoloquinoline derivatives **37** by heating 4-(arylidene)pyrazolin-5-one **36** and aromatic amine in the presence of zinc chloride.[24]

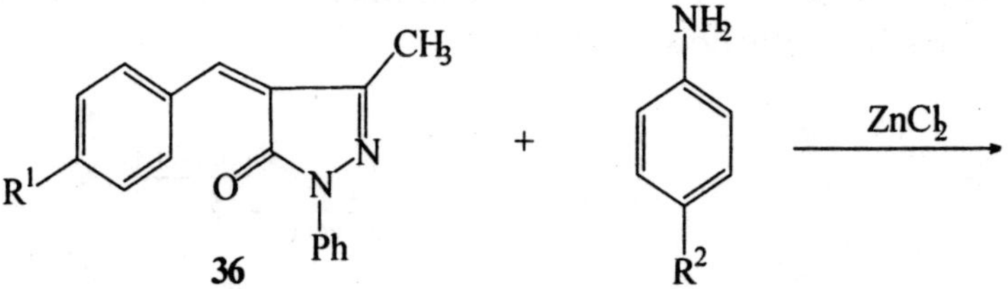

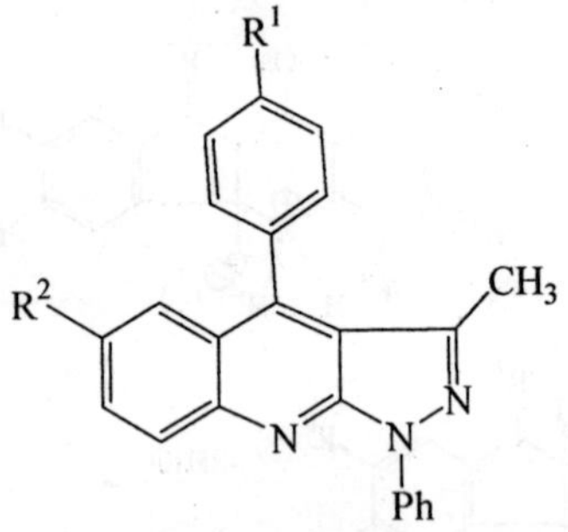

37

4. *Synthesis of quinoline derivatives using para toluene sulfonic acid under solvent free conditions:*

Recently, synthesis of quinoline has been reported under microwave solvent free condition in the presence of para toluene sulfonic acid as catalyst.[25]

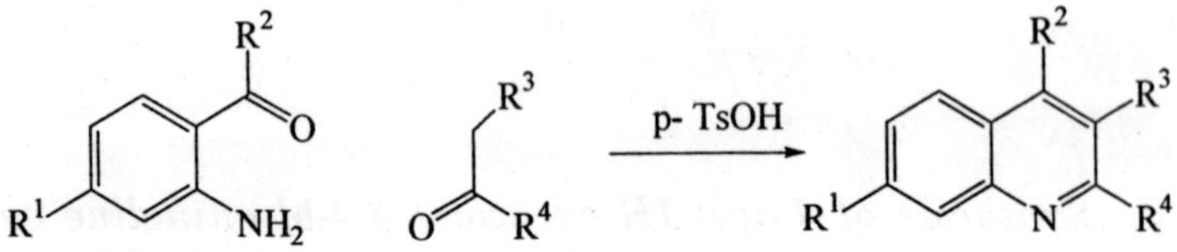

5. *Microwave assisted Friedlander quinoline synthesis from ketones and o-amino aryl ketones using silica sulfuric acid:*

An efficient green synthesis has been described for quinolines. The reaction involves microwave irradiation of a neat containing a mixture of o-amino ketone **38**, ketone **39** and silica sulfuric acid.[26]

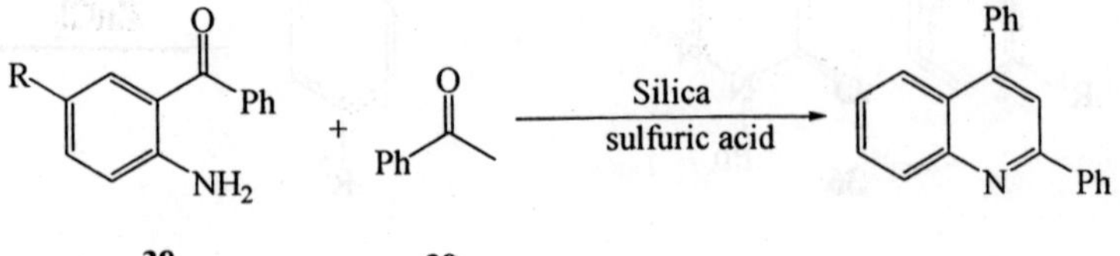

38 **39**

6. *Friedlander condensation in the presence of molecular iodine*

Quinoline derivatives have been also synthesized under mild conditions using molecular iodine as a new catalyst.[27] The method provides an excellent route to both quinolines, poly quinolines and avoids harsh reaction conditions and hazardous acids and bases.

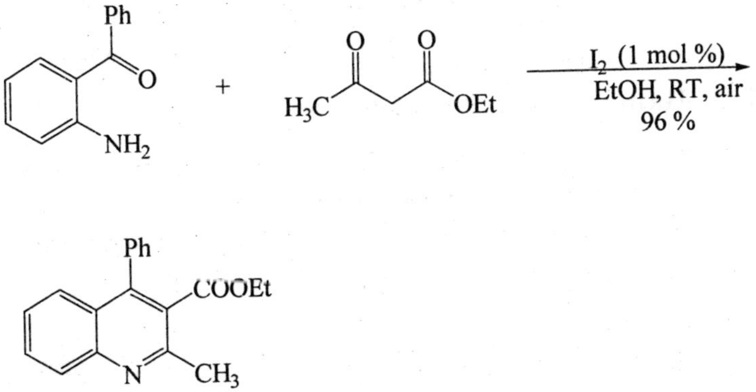

5.11. Gabriel synthesis

The reaction involves conversion of an alkyl halide to primary amine in the presence of potassium phthalimide[28,29] under basic conditions. In this reaction potassium phathalimide first reacts with alkyl halide to give N-alkyl phthalimide **40** which when subjected to hydrolysis affords primary amine and phthalic acid as by product.

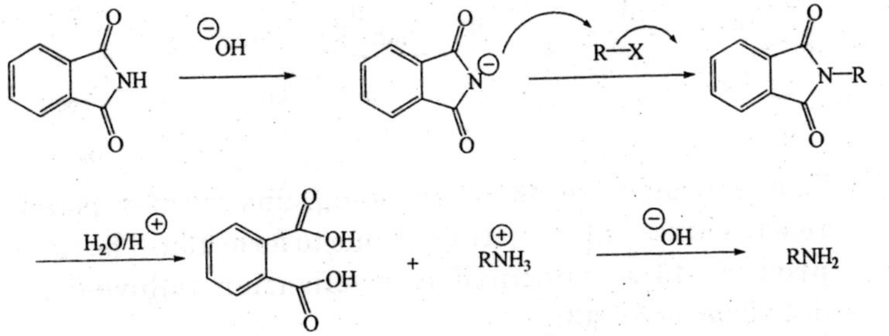

Benzoic anhydride has been converted to **41** by carrying out reaction with phenylacetic acid in the presence of potassium acetate under microwave irradiation condition.[30]

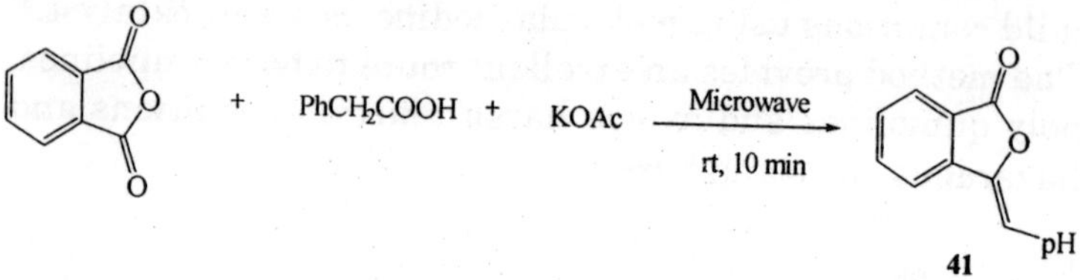

41

82%

5.12. Hantzch pyrrole synthesis

The reaction involves synthesis of pyrroles by carrying out reaction of active methylene compounds such as β-ketoester with α-chloro methyl ketones and ammonia/ primary amines.[31]

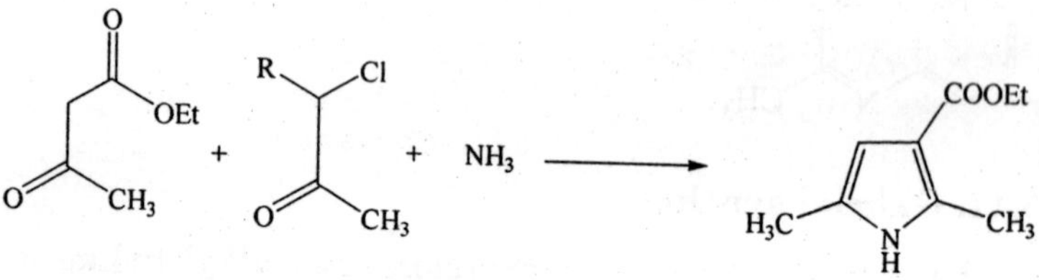

The rection involves following steps.

(i) Formation of intermediate **43** from the reaction of ethylacetoacetate **42** and ammonia.

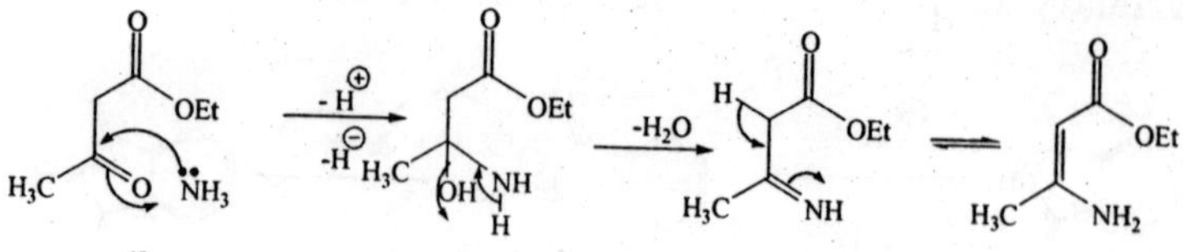

42 **43**

2. The intermediate **43** (β-amino-α,β-unsaturated ester) reacts with α-chloromethyl ketone to form **43a**. The final product **44** is obtained by cyclization followed by dehydration of **43a**.

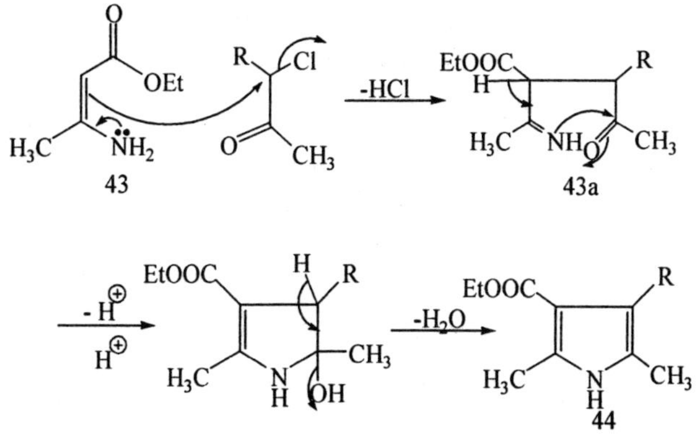

43 43a 44

5.13. Hinsberg oxindole synthesis

The reaction is used for the synthesis of oxindole and oxiquinoline derivatives.[32] The reaction takes place by treatment of secondary aryl amine and sodium bisulfite addition product of glyoxal **45**. The following steps are involved during the formation of the product.

1. Formation of addition product **45** by addition of sodium bisulfite to glyoxal.

$$NaHSO_3 \longrightarrow Na^{\oplus} + H^{\oplus} + SO_3^{2-}$$

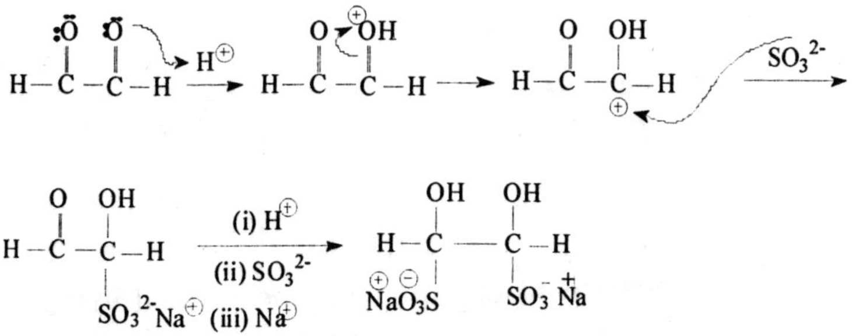

45

2. Interaction of amine with compound **45** takes place to form the intermediate **45a** which undergoes ring closure to form final product **45b**.

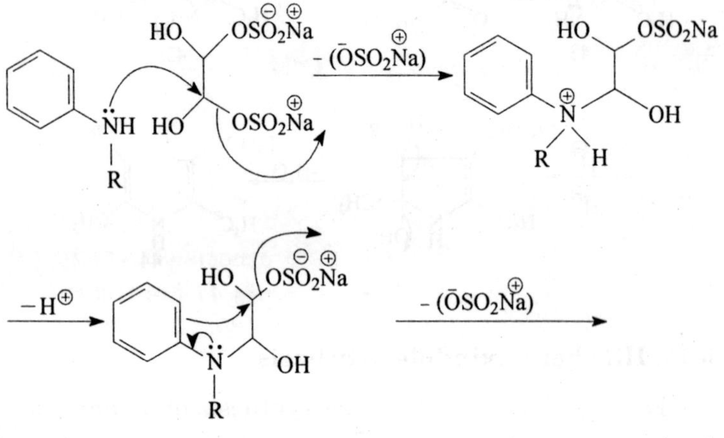

45a

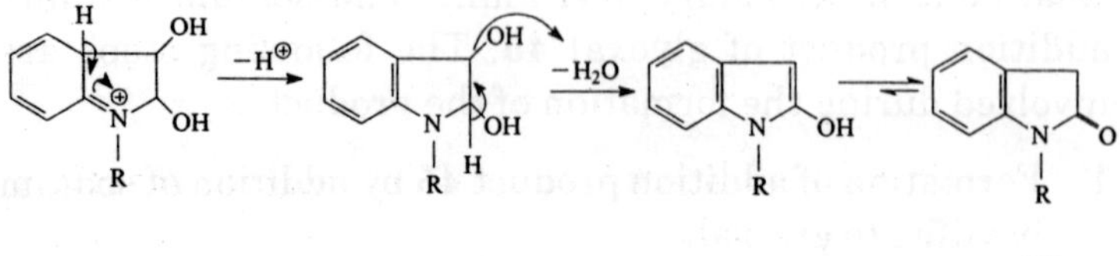

45 b

Oxindoles **46** are also synthesized from a haloacyl halide and secondary arylamines in similar manner[33]

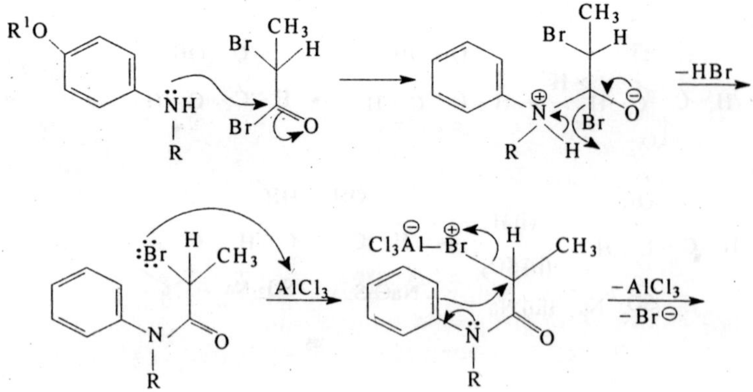

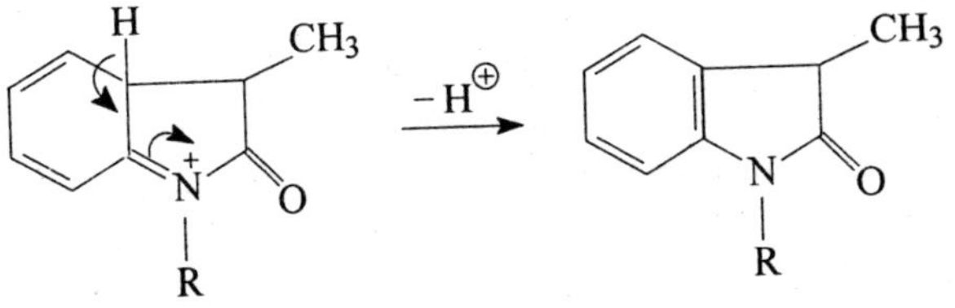

46

5.14. Knorr pyrazole synthesis

This reaction is used for the synthesis of pyrazoles **48**. The reaction occurs when β-diketones **47** are treated with hydrazines.[34]

The reaction takes place as shown below:

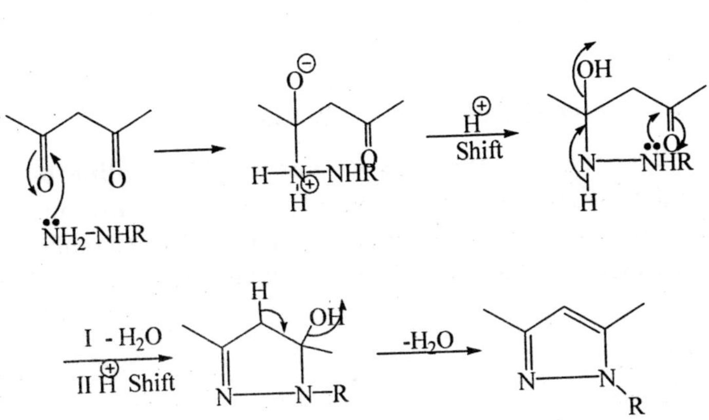

5.15. Knorr pyrrole synthesis

This reaction is also used for the synthesis of pyrrole derivative **51**.[35] The reaction takes place by the condensation of α-amino ketone **49** with carbonyl compounds having active methylene group **50** in the presence of an acidic or basic catalyst.

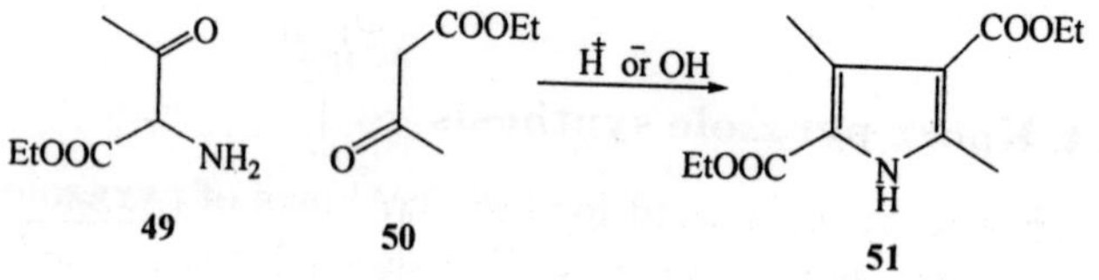

The mechanism of the reaction is outlined below:

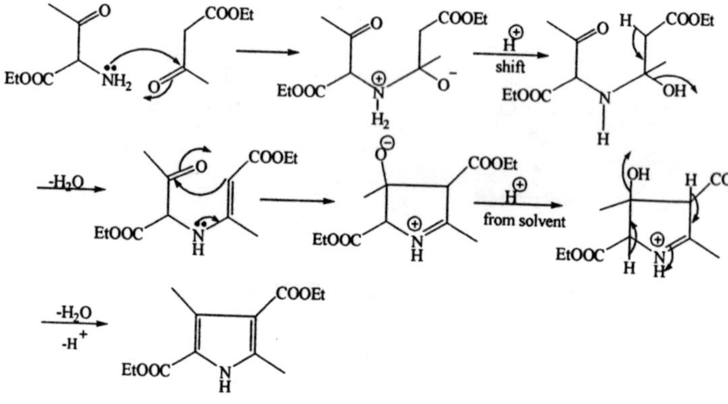

5.16. Knorr Quinoline synthesis

The reaction involves formation of quinoline by the condensation of a β-ketoester and primary aromatic amine.[36] The intermediate **52** which is formed during the reaction is cyclized to final product **53** upon heating with sulfuric acid.

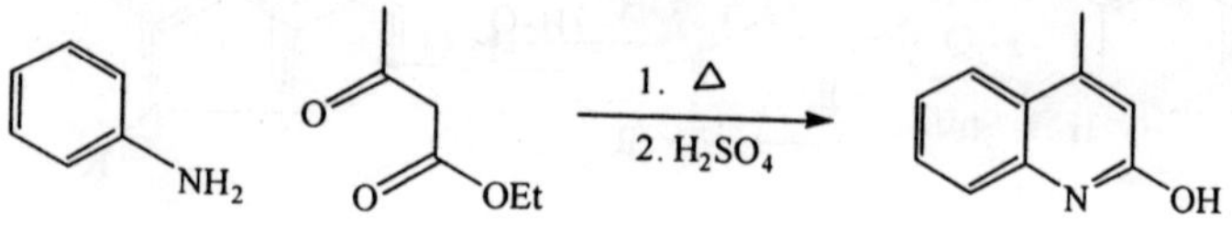

The reaction takes place as shown below:

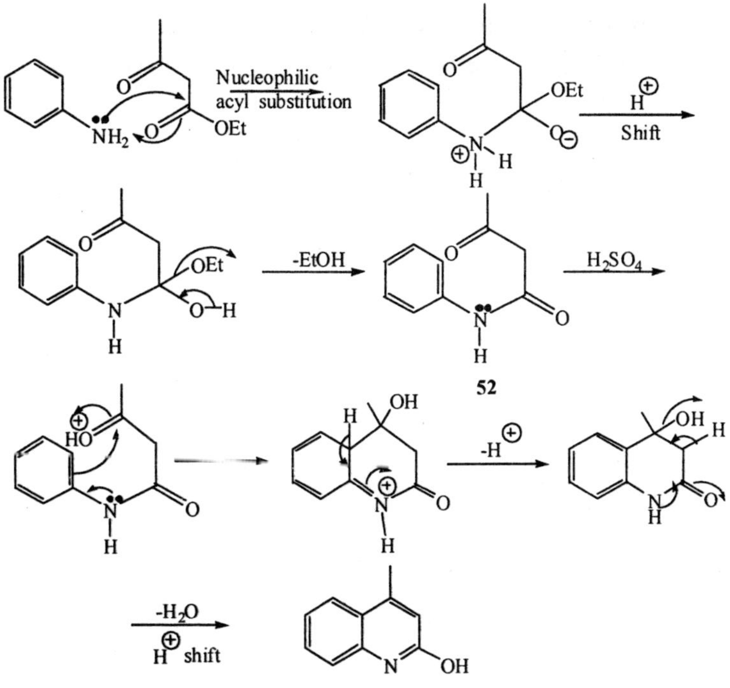

52

53

5.17. Niementowski Quinoline synthesis

The reaction involves formation of 4-hydroxyquinoline derivatives **54** from anthranilic acid and enolizable carbonyl compounds.[37]

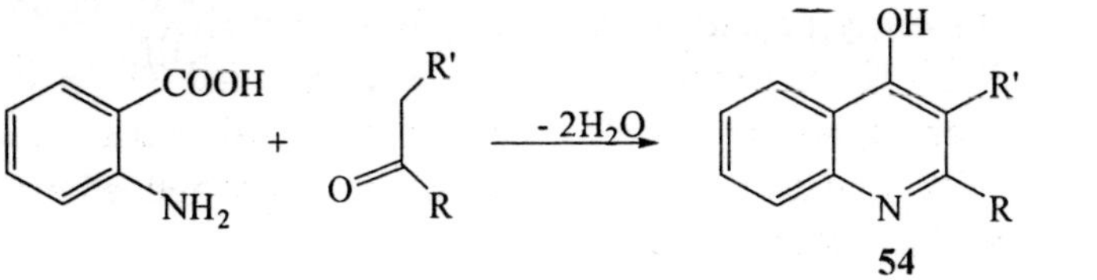

54

5.18. Paal-Knorr pyrrole synthesis

The reaction is used for the synthesis of pyrrole **56**. The formation of pyrrole takes place by heating a mixture of 1, 4-dicarbonyl compounds **55** with ammonia or ammonia derivatives such as primary amines and hydrazines.[38]

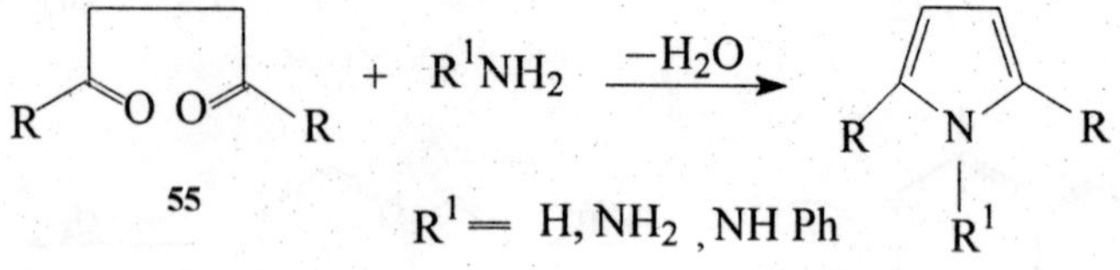

55

56

The condensation takes place as shown below:

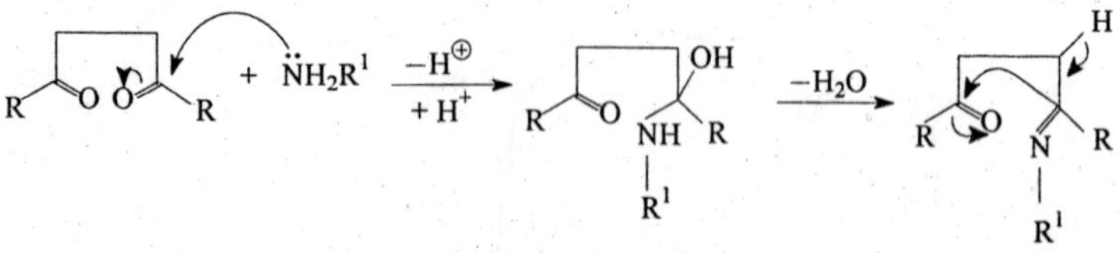

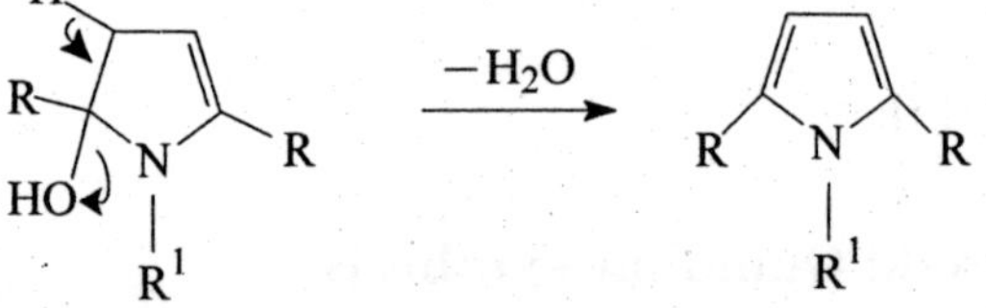

Recent examples of synthesis of pyrrole through Paal-Knorr Pyrrole synthesis:

1. Recently an easy synthesis of N-substituted pyrroles by Paal- Knorr condensation has been reported. The reaction is catalysed by scandium triflates and takes place under mild condition. The catalyst triflate can be reused further without loss of activity.[39]

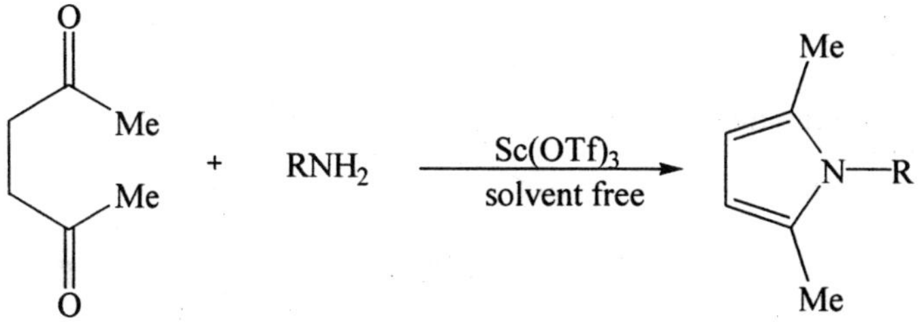

2. *Synthesis of 1,2,3,5,-tetrasubstituted pyrrole derivatives from bromoallyl 1,2-dicarbonyl compounds:*

Tetra substituted pyrrole derivatives **59** have been synthesized from 2-(2-bromoallyl-cyclohexane-1,3-dione in good yield.[40] The reaction involves alkylation of 1,3-dicarbonyl compound **57** to 2-(2-bromoallyl)-1,3-dicarbonyl compound **58**.

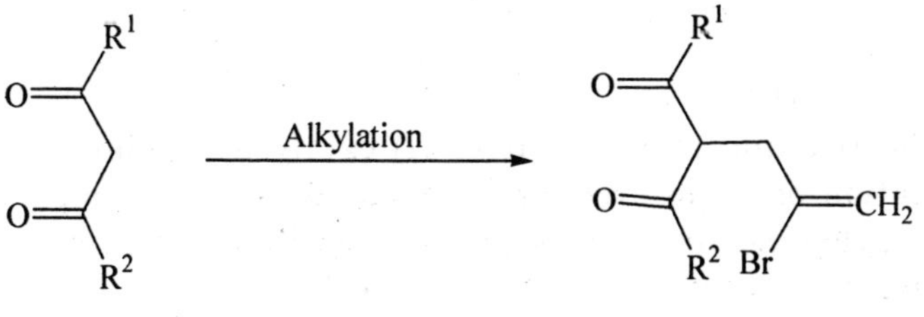

57 **58**

58 is then converted to the β-enamine **59a**. The base catalysed cyclization of **59a** affords pyrrole **59**.

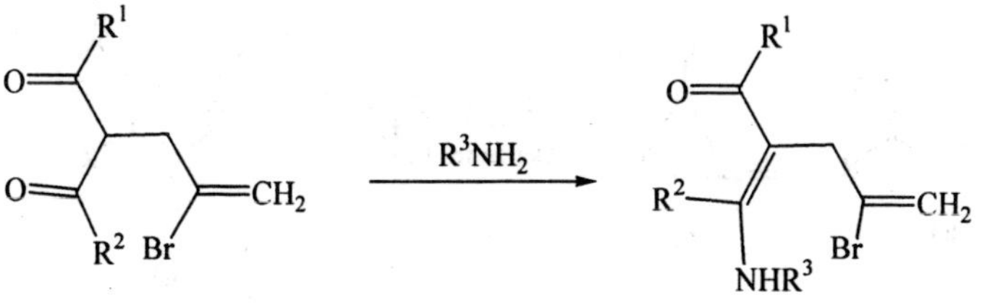

58 **59a**

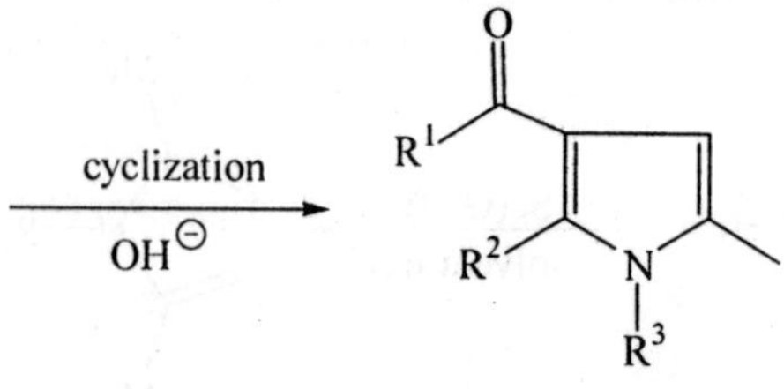

59

3. Recently, formation of a novel pyrrole via Paal-Knorr Synthesis has been reported in the presence of silver acetate at room temperature.[41]

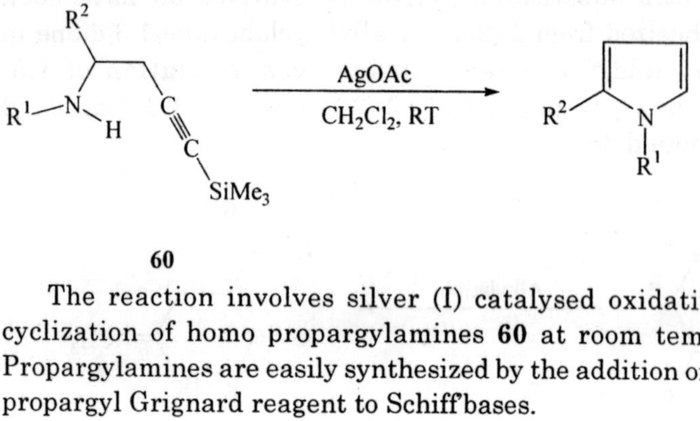

60

The reaction involves silver (I) catalysed oxidative cyclization of homo propargylamines **60** at room temp. Propargylamines are easily synthesized by the addition of a propargyl Grignard reagent to Schiff bases.

4. *Synthesis of 1,2,3,4-tetra-substituted pyrrole derivative*

Tetra substituted pyrrole derivatives **63** have been synthesized by the reaction of 1,3-diketone **61** with methylene azyridines **62** in the presence of palladium.[42]

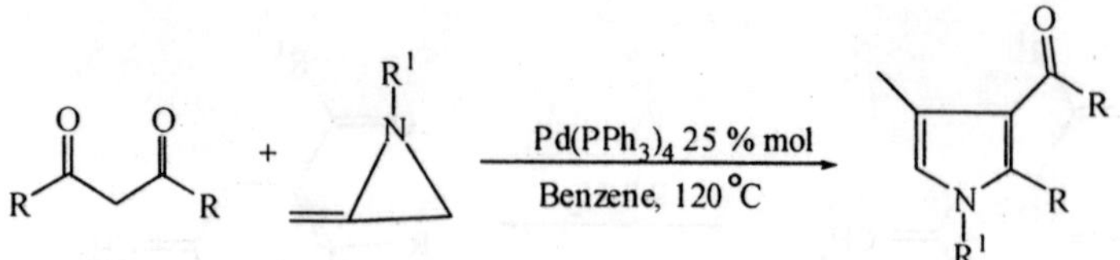

61 **62** **63**

5. Synthesis of pyrrole in the presence of sulfamic acid as a novel, efficient, cost effective and reusable solid acid catalyst under solvent free conditions has been also reported.[43] Thus, Paal- Knorr condensation of 2,5-hexadione with primary amines in the presence of sulfamic acid takes place with an excellent yield. N-substituted pyrrole derivatives, by this method, are obtained easily and rapidly.

5.19. Pictet- Spengler reaction

The reaction involves formation of tetrahydroisoquinoline derivative by condensation of β-aryl ethyl amines and carbonyl compounds, fol'owed by cyclization of Schiff's base formed.[44,45]

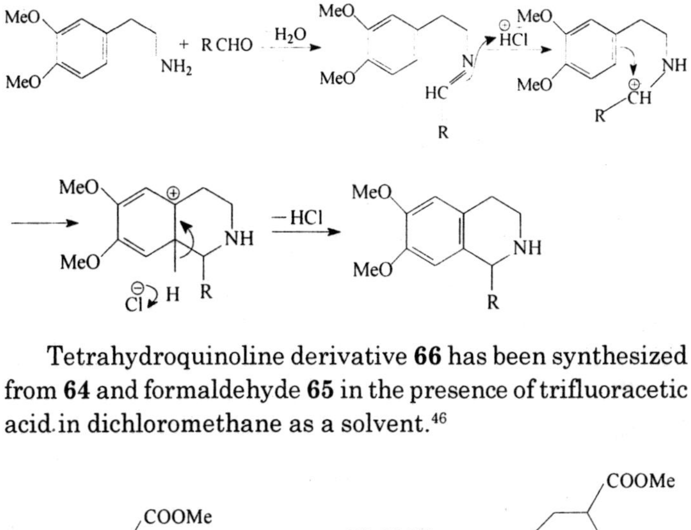

Tetrahydroquinoline derivative **66** has been synthesized from **64** and formaldehyde **65** in the presence of trifluoracetic acid in dichloromethane as a solvent.[46]

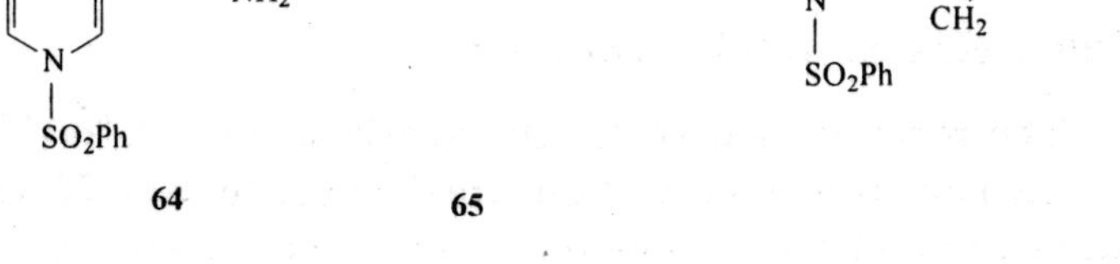

64 **65**

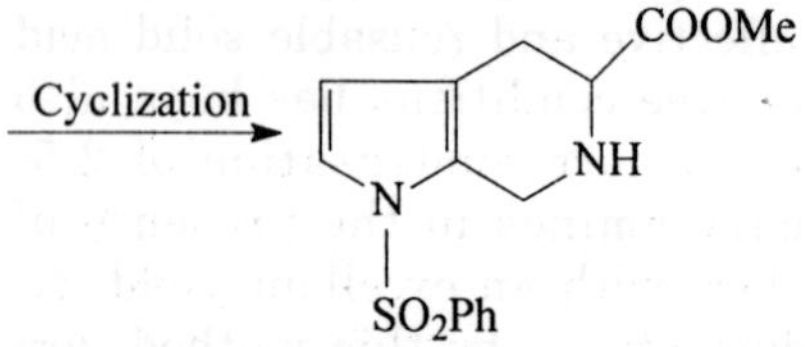

66

Synthesis of isoquinoline derivative **69** has been reported from β arylethyl amines **67** and ketone derivative **68** involving Pictet- Spengler followed by Diels Alder reaction using trifluroacetic acid as catalyst.[47]

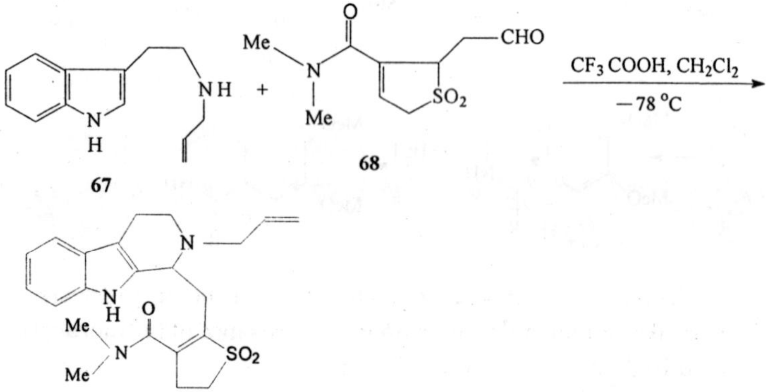

67 **68** **69**

5.20. Reissert indole synthesis

The reaction is used for the synthesis of indoles. The reaction involves condensation of ortho nitro toluene **70** with oxalic ester **71** followed by reduction and cyclization of the intermediate **72** to the final product indole **73**.[48,49]

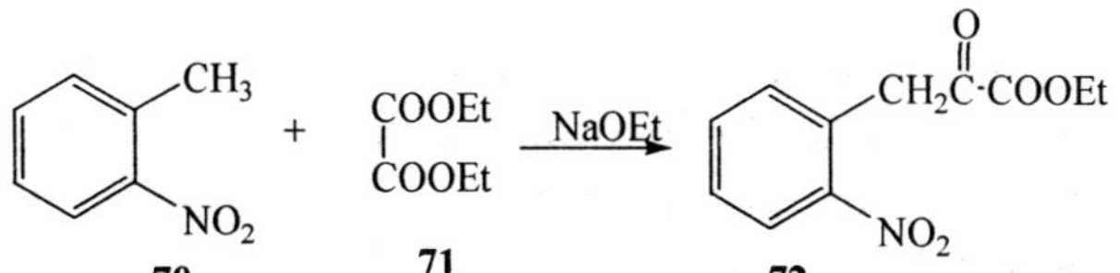

70 **71** **72**

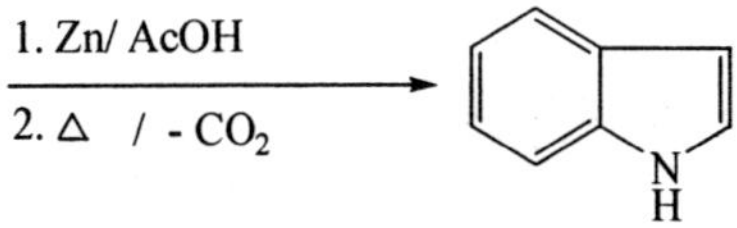

73

The mechanism of the reaction is outlined below:

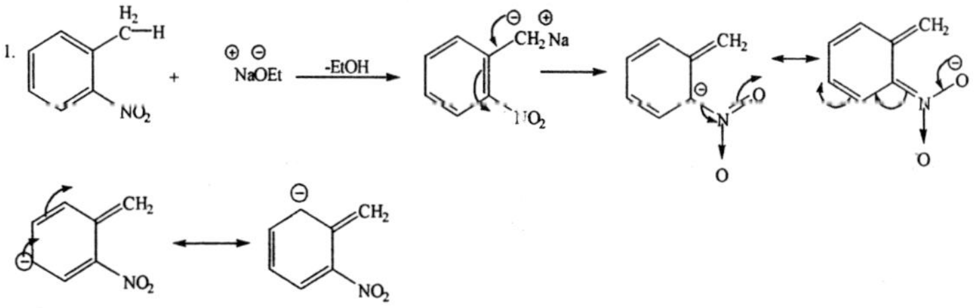

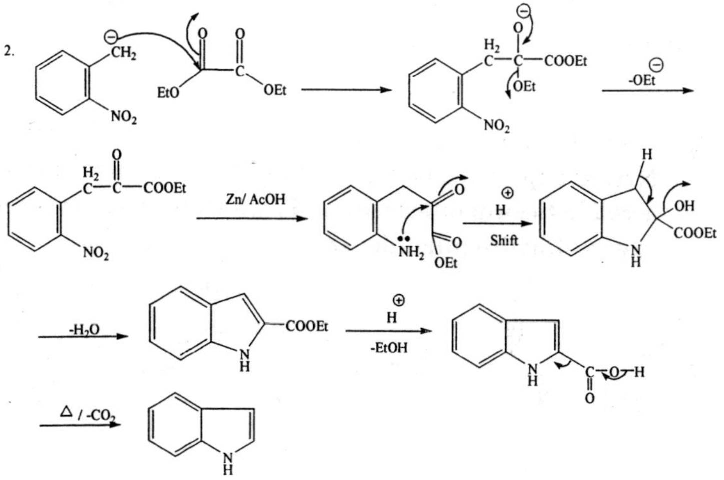

5.21. Skraup synthesis

This reaction is used for the synthesis of quinolines.[50] The reaction takes place by heating a mixture of aromatic amine, glycerol, an oxidizing agent (e.g. nitrobenzene) in the presence of sulfuric acid.

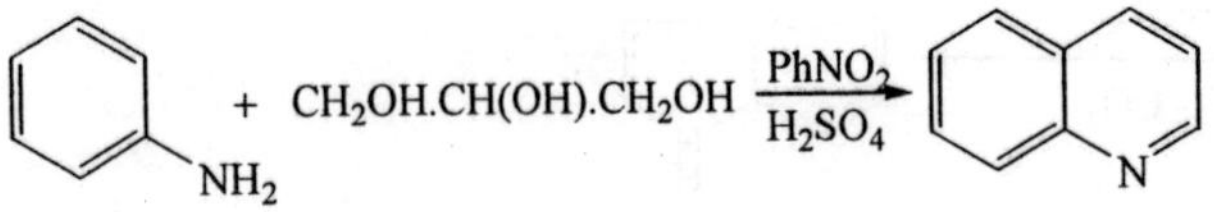

The mechanism of the reaction involves following steps.

1. Formation of acrolein **75** by dehydration of glycerol **74** with concentrated sulfuric acid.

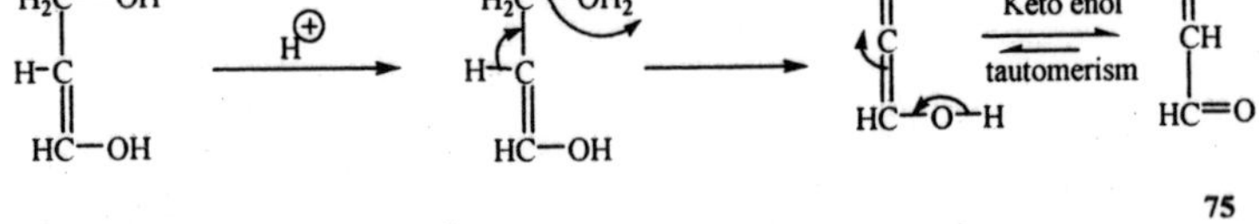

74, 75

2. Michael addition takes places by nucleophilic attack of NH_2 group of aromatic amine **76** on double bond of acrolein to form intermediate **77** which is finally cyclised to give dihydroquinoline **78**.

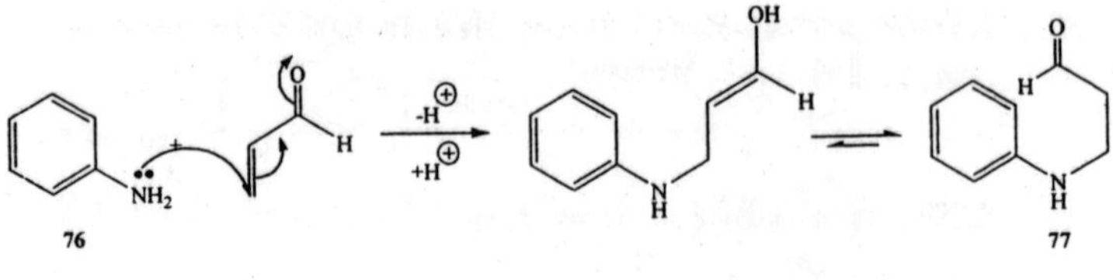

76, 77

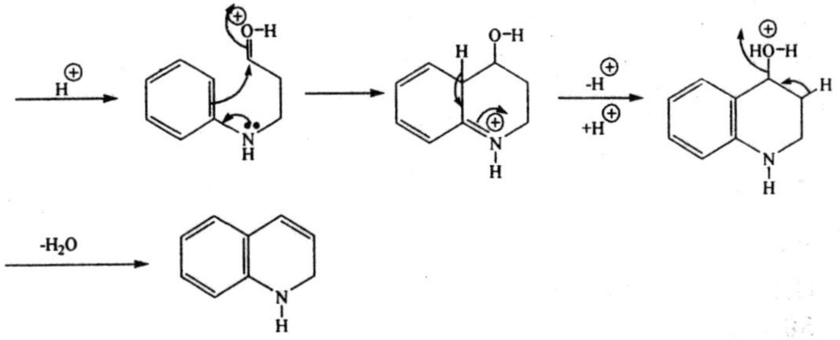

78

3. Oxidation of dihydroquinoline **78** by nitrobenzene affords quinoline **79**.

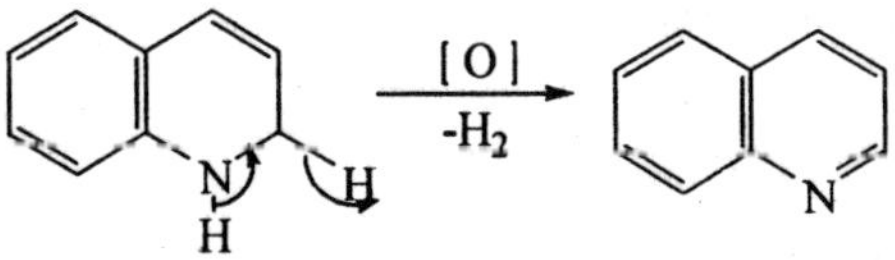

78 **79**

REFERENCES

1. Mayo, P. De, *Molecular Rearrangement*, Vol.II, Interscience Publishers, 1964.
2. H. S. Isbell and H. L. Frush, *J. Org. Chem.*, 1958, **23**, 1309.
3. J. E. Hodge and C. E. Rist, *J. Am. Chem. Soc.*, 1952, **74**, 1494, 1952, 75, 316,.
4. L. Rosen, J. W. Woods, and W. Pigman, *Chem. Ber.* 1957, **90**, 1038,.
5. A. Bischler and B. Napieralski, *Ber.*, 1903,1893, 26, 1891,.
6. J. Quirante, C. Escolano, A. Merino and J. Bonjoch, *J. Org. Chem.*, 1998, **63**, 968.
7. P. A. Wender and T. E. Smith, *J. Org. Chem.*,1996, **61**, 824.
8. A. E. Chichibabin, *J. Russ. Phys. Chem. Soc.*, 1906, **37**, 1229; *J. Prakt. Chem.*, 1924, 107, 122

9. A. E. Chichibabin and O. A. Zeide, *J. Russ. Phys. Chem. Soc.*, 1914, **46**, 1216: *Chem . Abstr.*, 1915, 9, 1901

10. L. Panzella, P. D. Donato, S. Comes, A. Napolitano, A. Palumbo and M. d'lschia, *Tetrahedron Lett.*, 2005, **46(38),** 6457.

11. A. Combes, *Bull. Soc. Chim. France*, 1888, **49**, 89.

12. O. Deobner and W. v. Miller, *Ber.* , 1881,**14**, 2812.

13. G. A. Dauphinee and T. P. Forrest, *Canadian J. Chem.*, 1978, **56**, 632.

14. G. Siva Prasad, R. Rajesh and P.T. Perumal, *Tetrahedron Lett.* , 2006, **47(11)**, 1783.

15. E. Fischer, F. Jourdan, *Ber.* , 1883, 16, 2241.

16. S. Wagaw, B. H. Yang and S. L. Buchwald, *J. Am. Chem. Soc.*, 1998, **120**, 6621.

17. D. Xu , W. Yang, S. Luo, B. Wang, J. Wu and Z. Xu, *Eur. J. Org. Chem.*, **2007,** (6), 1007.

18. S. M. Hutchins and K. T. Chapman, *Tetrahedron Lett.* , 1996, **37 (28)**, 4869.

19. O. Miyata, Y. Kimura, K. Muroya, H. Hiramatsu, and T. Naito, *Tetrahedron Lett.* , 1999, **40**, 3601.

20. A. Hantzch, *Ber.* , 1885, **18**, 1744.

21. P. Friedlander, *Ber.* , 1882, **15**, 2572,.

22. G. Sabitha, R. S. Babu, B. V. S. Reddy and J. S. Yadav, *Synth. Commun.* , 1999, **29(24),** 4403.

23. F. Xu, D. Q. Yang, K. L. Jiyang and W. Guo, *Chinese. Chem. Lett.*, 2006, **17 (2),** 187.

24. K. Chaczatrian, G. Chaczatrian, A. Danel and P. Tomasik, *Arkivoc*, 2001 (vi), **63**.

25. G. W. Wang, *Org. Biomol. Chem.*, 2006, **4,** 104.

26. M. A. Zolfigol, P. Salehi, M. Shiri, T.F. Rastegar and A. Ghaderi, *J Iran. Chem. Soc.* , 2008, **5 (3),** 490.

27. J. Wu, H. G. Xia and K. Gao, *Org. Biomol. Chem.*, 2006, **4**, 126.

28. S. Gabriel, *Ber.* , 1887, **20**,2224.

29. M. S. Gibson, R. W. Bradshaw, *Angew. Chem. Int. Ed, Engl.* , 1968, **7**, 919.

30. M. Lacova, J. Chovancova, E. Veverkova and S. Toma, *Tetrahedron*, 1996, **52**, 14995.

31. A. Hantzsch, *Ber.* , 1890, **23**, 1474.

32. O. Hinsberg , *Ber.* , 1885, **21**, 110; 1892**, 25,** ,45.

33. R. Stolle, *Chem. Ber.*, 1913, **46**, 3915; W. Sumter, *Chem. Rev.* , 1994, 34, 396.

34. L. Knorr, *Ber.* , 1883, **16**, 2587.

35. L. Knorr, *Ber.* , 1884, **17**, 1635.

36. L. Knorr, *Ann.* , 1886,236, **69** ;1888), 245, 375, 378.

37. S. V. Niementowski, *Ber.* , **27**, 1394 (1894) ; *ibid*, **28**, 2809 (18195).

38. C. Paal, *Ber.* , 1885, **18,** 367; L. Knorr, *ibid.* , 299.

39. J. Chen, H. Wu, Z. Zheng, C. Jin, X. Zhang and W, Su, *Tetrahedron Lett.* , 2006, **47 (30),** 5383.

40. A. S. Demir, I. M. Akhmedov and O. Sesenoglu, *Tetrahedron*, 2002, **58 (49),** 9793.

41. S. Agarwal and H. J. Knolker, *Org. biomol. Chem.*, 2004, **2**, 3060.

42. K. A. D. S. Kathriarachchi, A. I. Siriwardana, I. Nakamura and Y. Yamamoto, *Tetrahedron Lett.* , 2007, **48 (13),** 2267.

43. S.K. De, *Synth. commun.*, 2008, **38 (5),** 803-809.

44. A. Pictet and T. Spengler, *Ber.* , 1911, **44**, 2030.

45. W.M. Whaley and T.R. Govindachari, *Org. React.* , 1951, **6,** 74

46. J.F. Rousseau and R.H. dodd, *J. Org. Chem.* , 1998, **63**, 2731.

47. J. Leonard, A.B. Hague and M.F. Jones, *Tetrahedron Lett.*, 1997, **38**. 3071.

48. A. Reissert, *Ber.*, 1897, **30**, 1030,.

49. J. G. Cannon et. al. *J. Med. Chem* ., 1981, **24**, 238,

50. Z. H. Skraup, *Ber.* , 1880, **13**, 2086.

6

Synthesis of Aromatic Compounds

6.1. Baeyer Drewson indigo synthesis

By this route indigo **1** is prepared involving aldol condensation [1]

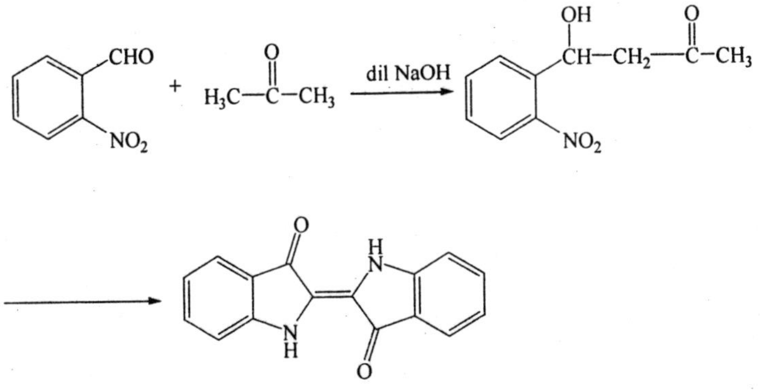

1

6.2. Biginelli Reaction

Synthesis of 3,4-dihydropyrimidine-2(H)-ones **2** by one

pot condensation reaction of an aromatic aldehyde, urea and ethyl acetoacetate in the presence of an acid catalyst is called as Biginelli reaction after the name of the scientist P. Biginelli who explored this reaction in 1819.

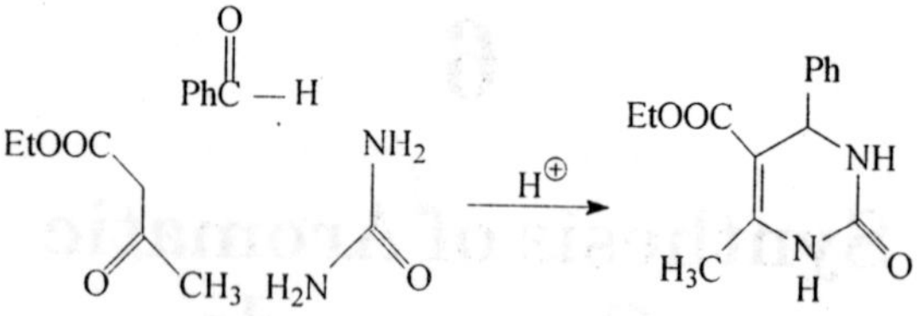

2

The classical Biginelli reaction affords poor yields when aldehydes used are aliphatic and or having hindered carbonyl group due to presence of ortho substitutents (e.g. salicyladehyde). Thus, the reaction was slightly modified by Atwal et al., which involves formation of α, β–unsaturated carbonyl compound **3** as intermediate which then condenses with modified urea in basic medium.[2-4]

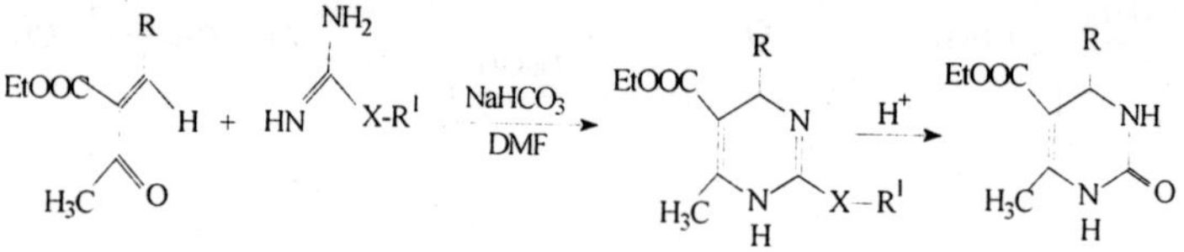

3

R' = Me, 4-methoxybenzyl

In addition to these other methods that lead to Biginelli compound are summarized below.

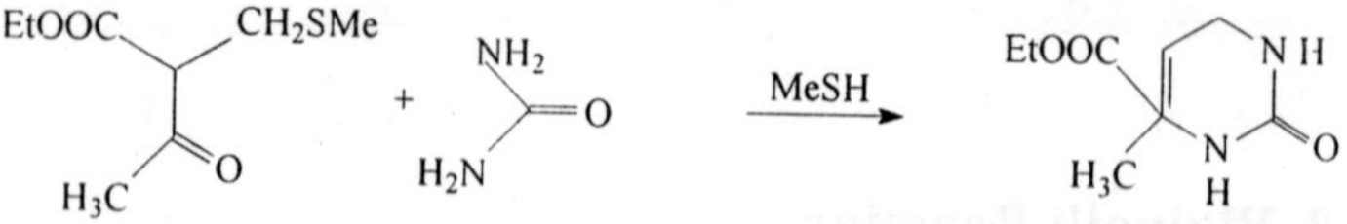

(Ref.5

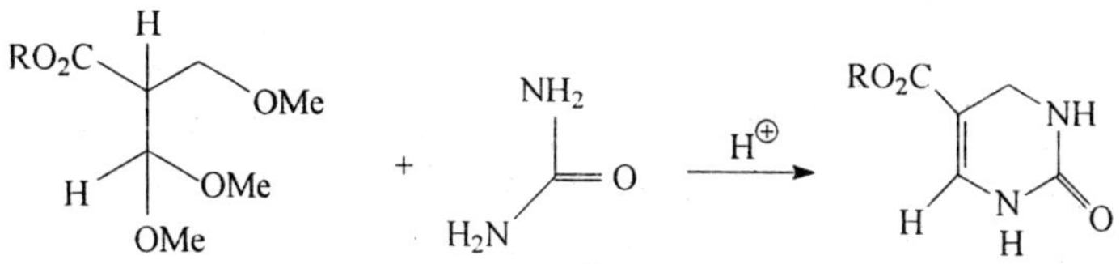

(Ref.6,7)

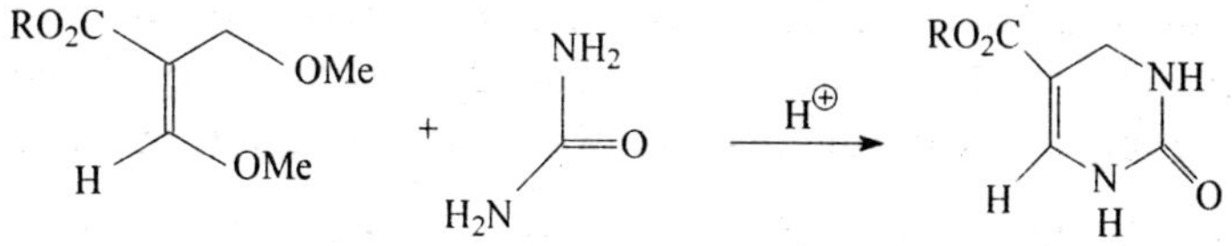

(Ref.7)

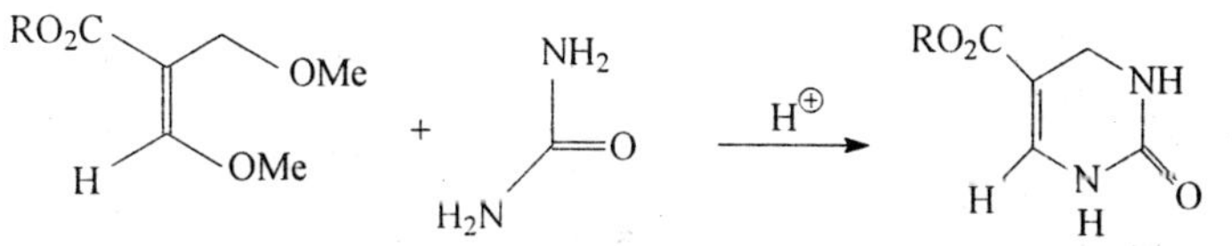

(Ref.7,8)

Functionalization of Biginelli Compounds

1. Biginelli compounds have been easily alkylated through nucleophilic substitution reaction with alkyl or acyl halides on S and N centres e.g.

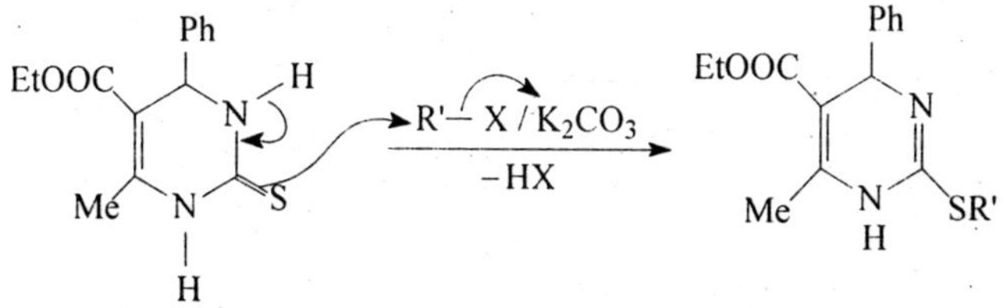

R' = H, Me

(Ref. 9)

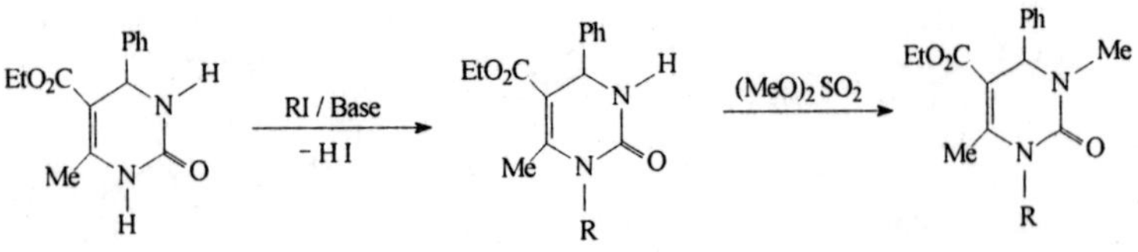

(Ref. 10-12)

In most of the cases regiospecific N-3 alkylation occurs but sometimes the reaction affords a mixture of N-1 and N-3 alkylated products also. The regioselectivity depends strongly on the substrate as well as the alkylating reagent.

2. Acylation of Biginelli compound takes place with Ac_2O whereas formylation is possible with DMF / $POCl_3$ at N-3[13]

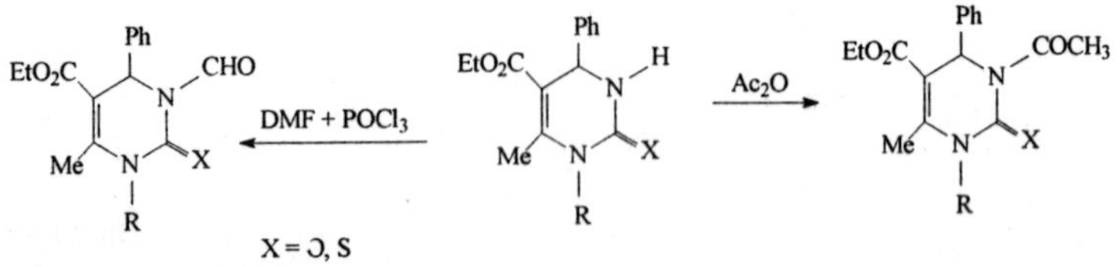

3. Ethoxy carbonylation at N-3 in biginelli compounds takes place with ethylchloroformate in the presence of NaH.[14]

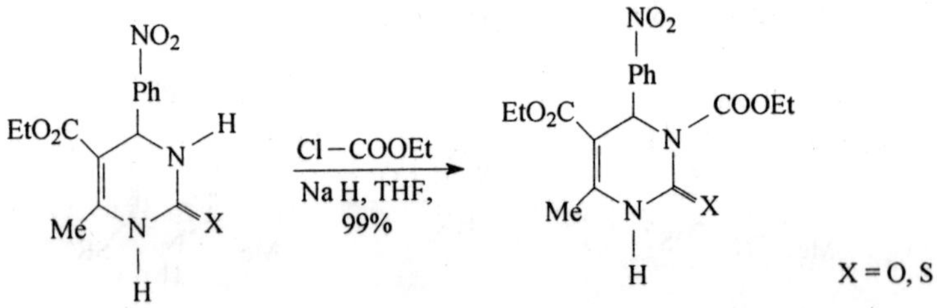

4. Ester group in Biginelli compounds is rather inert. However, basic hydrolysis converts ester group into carboxyl group[15], which can be decarboxylated to form 5-unsubstituted dihydro pyrimidine 4 derivative.

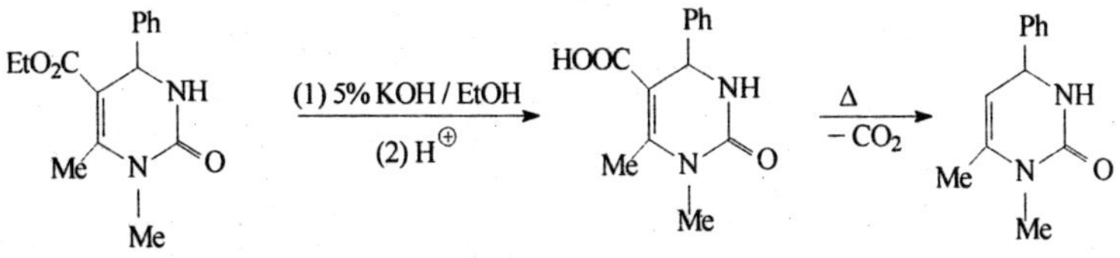

4

5. Dihydropyrimidine containing COOH group at 5-position **5** can be converted to azide derivative **6** which in turn can be converted into isocyanides **7** through Curtius rearrangement[16]

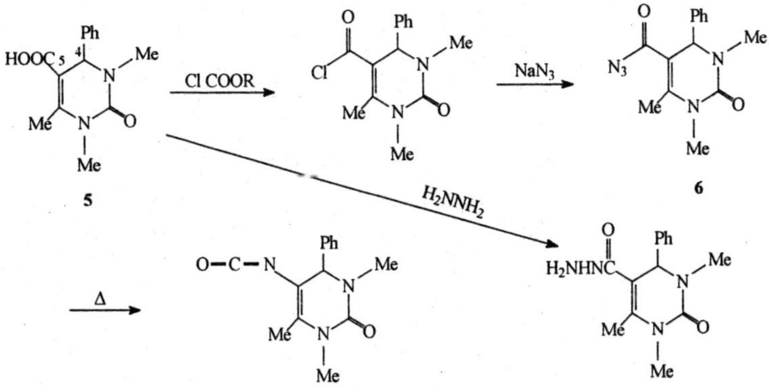

5 6 7 8

Further COOH can be transformed into hydrazide **8** by treatment with hydrazine[17]

Some nucleophilic substitutions reactions are also possible as indicated below:[18]

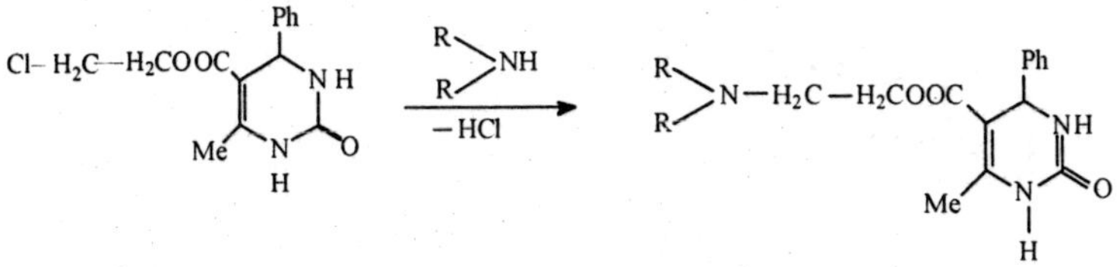

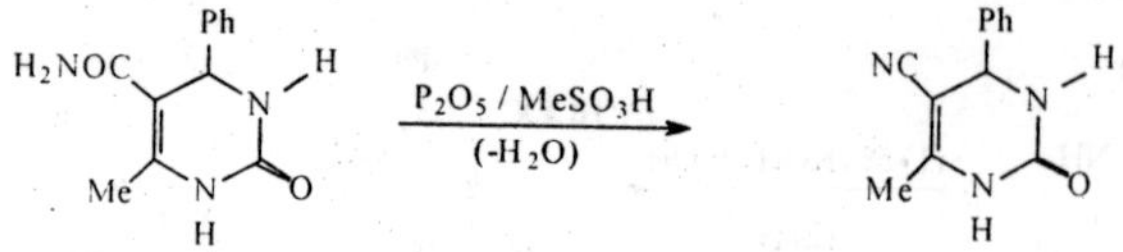

6. Functionlization at C-6 methyl group has been also reported. Thus, bromination at room temperature gives bromo compound which can be transformed in to other synthetically important intermediates.[19,20]

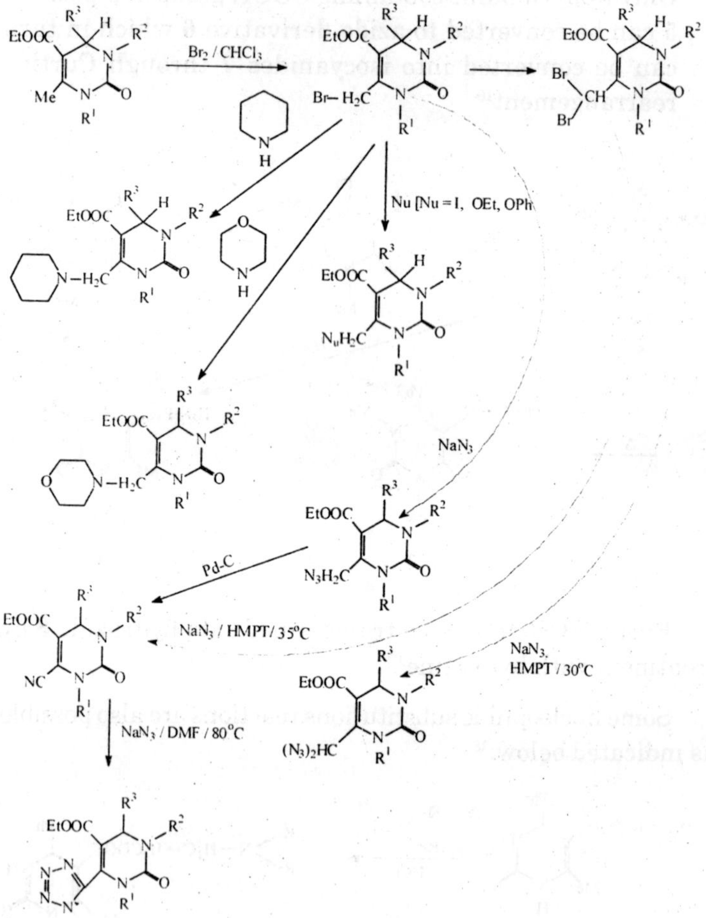

7. Cyclization reaction involving C-6 and C-5 carbon centers are also reported by the researchers. Thus, furopyrimidine **10** is obtained upon thermolysis of **9**.[21]

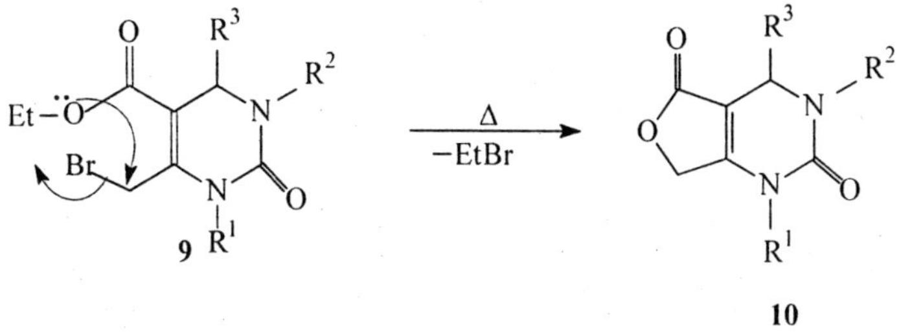

Transformation of dihydropyrimidine to other fused pyrimidine derivatives are summarized below.[22,23]

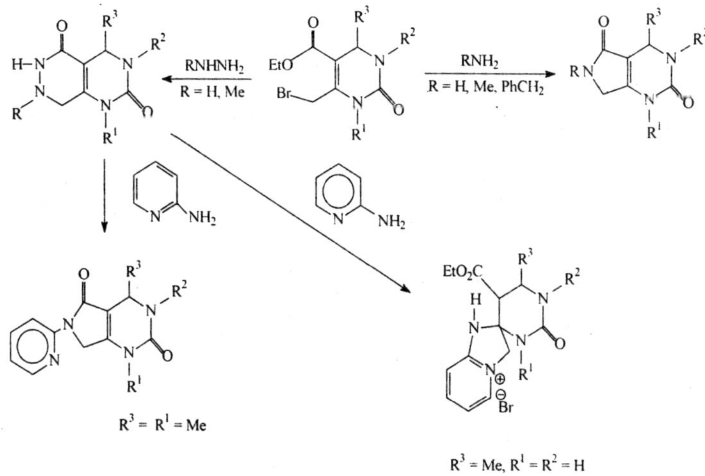

6.3. Pechmann condensation

The reaction is used for the synthesis of coumarins. The reaction takes place by condensation of phenol with ethylacetoacetate in the presence of a Lewis acid.[24]

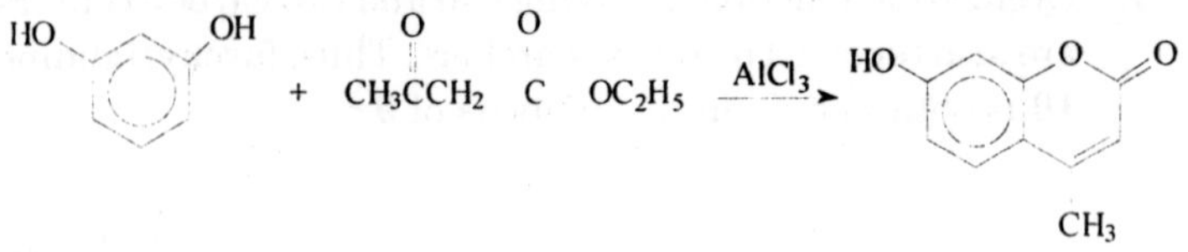

The mechanism of the reaction involves following steps:

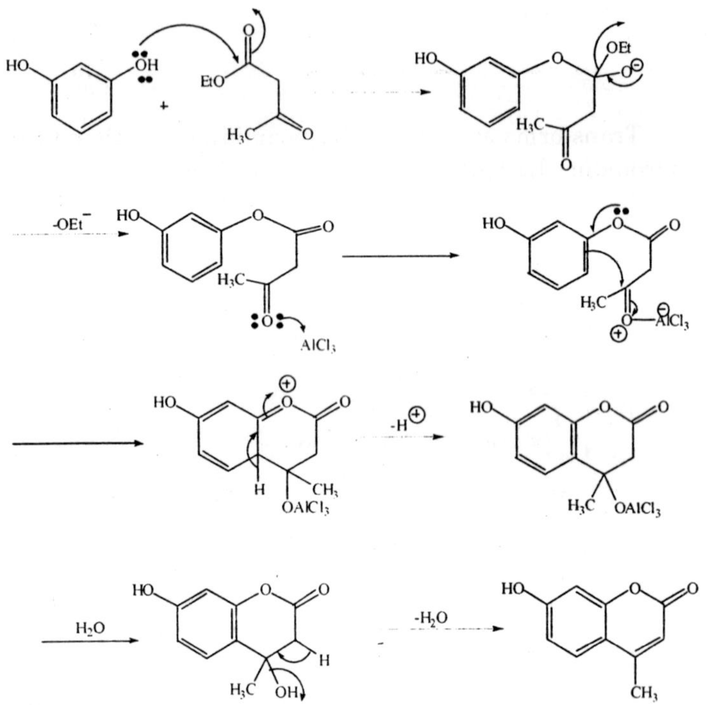

Protonic acids have been also used as condensing agents.Thus,a mixture of coumarins and chromones was obtained by condensation of 2,7-dihydroxy naphthalene and ethylacetoacetate in the presence of sulfuric acid.[32]

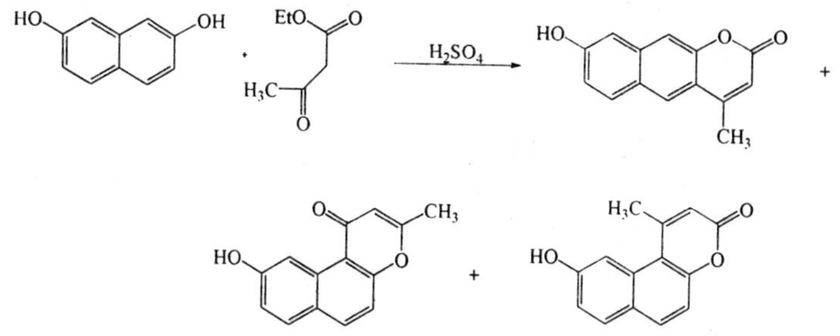

Besides protic acids other catalysts have been also employed for the synthesis of coumarins e.g. $ZrCl_4$ has been used as condensing agent in the synthesis of coumarins from phenol and ethylacetoacetate under solvent free conditions.[33]

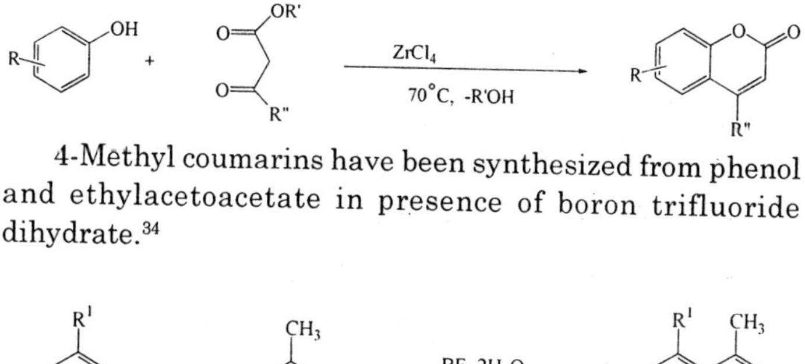

4-Methyl coumarins have been synthesized from phenol and ethylacetoacetate in presence of boron trifluoride dihydrate.[34]

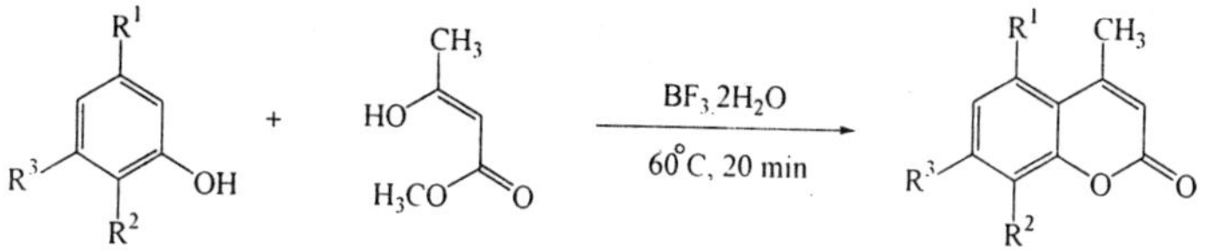

Recently synthesis of 4-substituted coumarins has been reported from phenol and ethylacetoacetate employing lithium bromide as catalyst under solvent free conditions.[35]

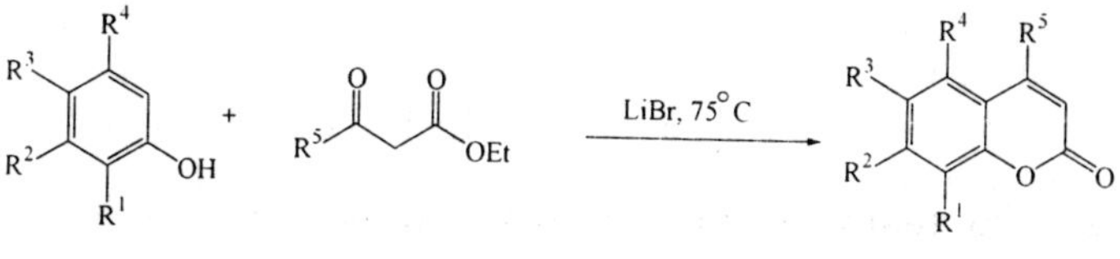

6.4. Simmonis chromone synthesis

The reaction is used for the synthesis of benzopyran-4-one (chromone).[25] Formation of chromone **11** takes place by heating a mixture of phenol, β-ketoester in the presence of a dehydrating agent such as phosphorous pentoxide or phosphorous oxychloride or concentrated sulfuric acid.

A small quantity of benzopyran-2-one (coumarin) **12** is also formed during the reaction.

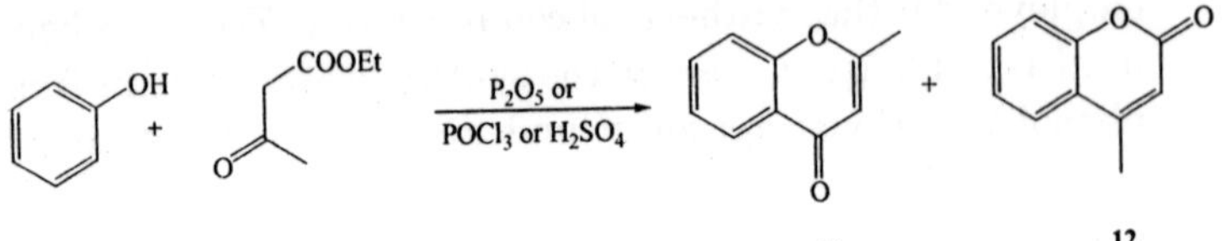

In the presence of concentrated sulfuric acid the reaction takes place as shown below:

A. **Formation of chromone**

1. Formation of carbocation **13** takes place by protonation of ethoxy group of ethylacetoacetate followed by removal of a molecule of EtOH

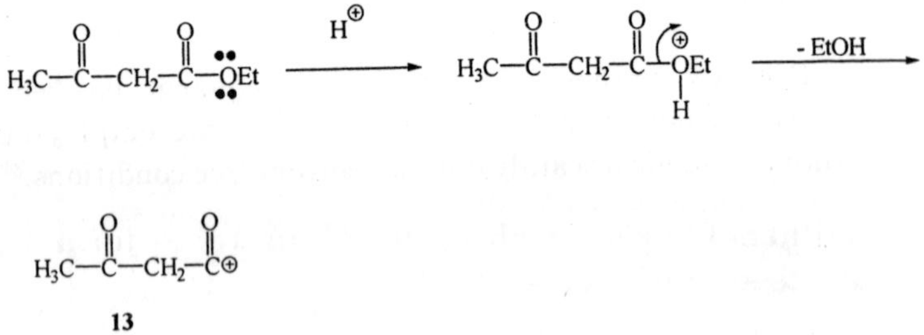

2. The reaction of phenol with carbocation **13** takes place to form β-diketone **14**

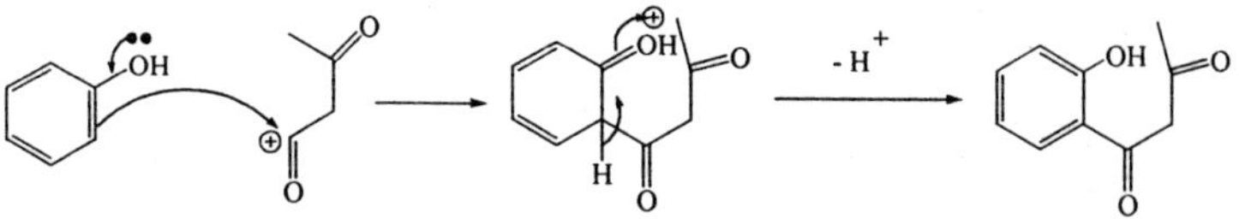

14

3. Intramolecular cyclization of β-diketone **14** occurs which is followed by dehydration to give 2-methylchromone **11**

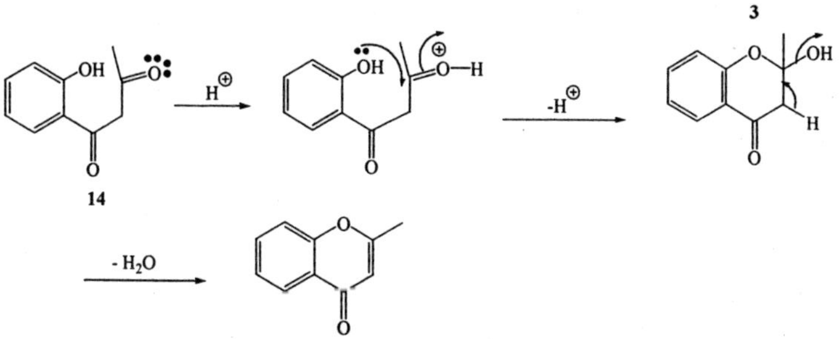

11

B. **Formation of coumarin**

Formation of coumarin involves following steps

1. Formation of carbocation **15** takes place by protonation of carbonyl group of ethylacetoacetate

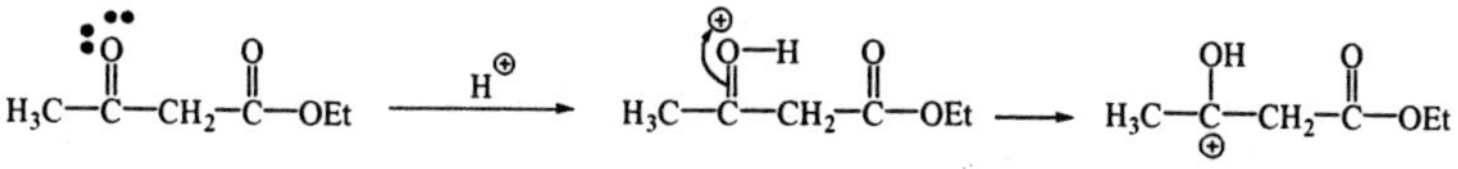

15

2. Phenol reacts with carbocation **15** to form α,β-unsaturated ester **16**.

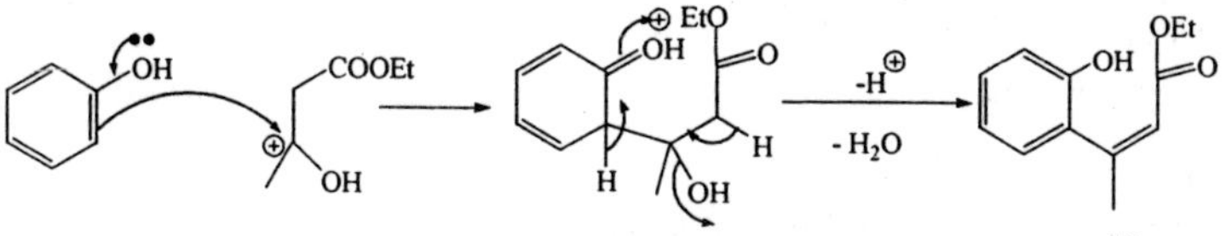

16

3. Cyclization of **16** takes place to form coumarin **12**.

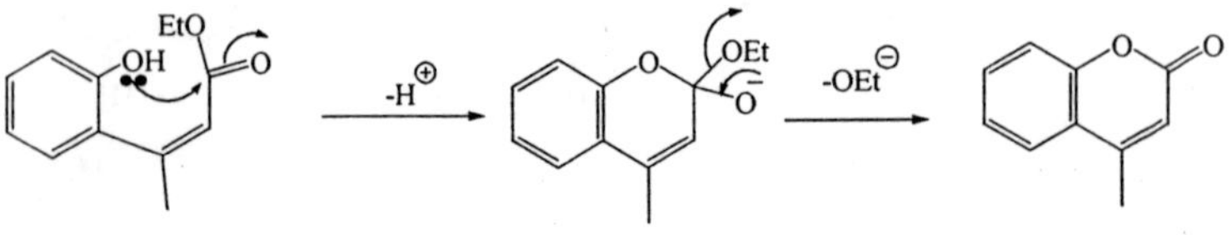

16 **12**

6.5. Suzuki Coupling

The reaction involves coupling of organic halides or perfluorinated sulfonates with organoboron derivatives. The reaction is catalyzed by palladium and proceeds with high stereo and regioselectivity.[26]

$$R^1 - BZ_2 + R^2 - Y \xrightarrow[\text{Base}]{[\text{Pd}]} R^1 - R^2$$

BZ_2 : $B(OR)_2$, 9 – BBN (Borabicyclononane) etc.

X : I, Br, Cl etc.
[Pd] : Pd $(PPh_3)_4$ etc.

Thus, e.g. alkene **14** has been synthesized from triflate **13** in quantitative yield.[27]

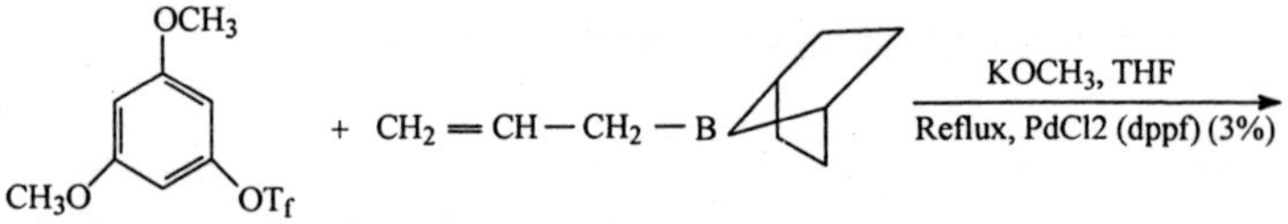

13

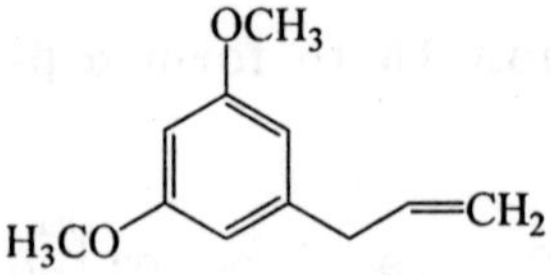

14

85%
dppf = diphenylphosphinoferocene.

6.6. Wharton reaction

The reaction is used for the synthesis of allylic alcohols from α, β-epoxy ketones. The reaction consists in treating a, β-epoxy ketones with hydrazine, when reduction takes place to afford allylic alcohols.[28]

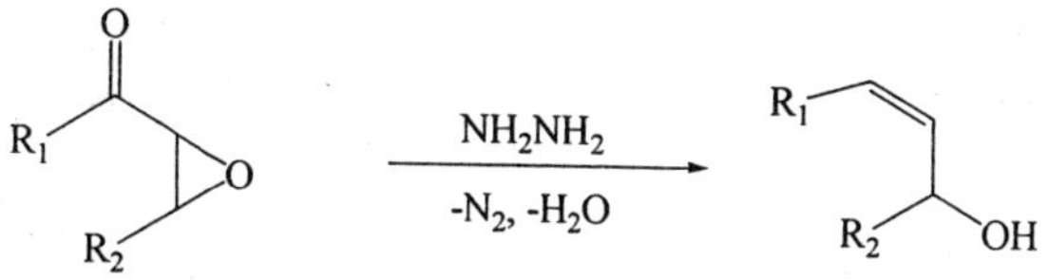

6.7. Wurtz reaction

The reaction is used for the synthesis of alkanes. The reaction takes place by coupling of alkyl halide with sodium.

When a mixture of two different alkyl halides is used, a mixture of alkanes is obtained which is difficult to separate.

In another view there is formation of intermediate organo sodium compound which undergoes nucleophilic substitutipon with alkyl halides to form hydrocarbon

6.8. Wurtz-Fittig reaction

The reaction is used for the synthesis of alkylated aromatic hydrocarbons. The coupling reaction takes place when alkyl halide and aryl halide are mixed with sodium in dry diethyl ether.

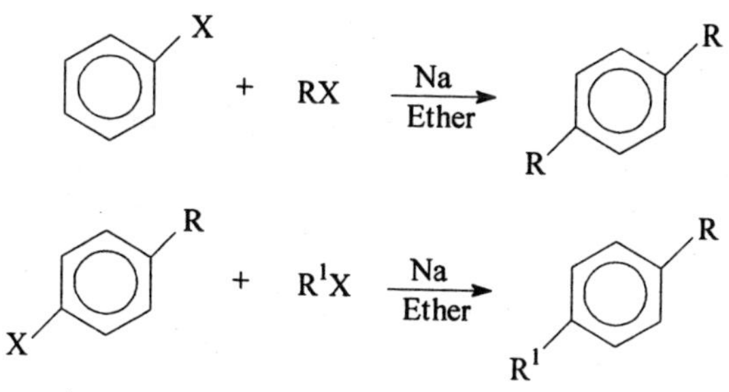

The reaction proceeds via free radical pathway and involves following steps.

1. Formation of free radicals in the presence of sodium in diethyl ether takes place.

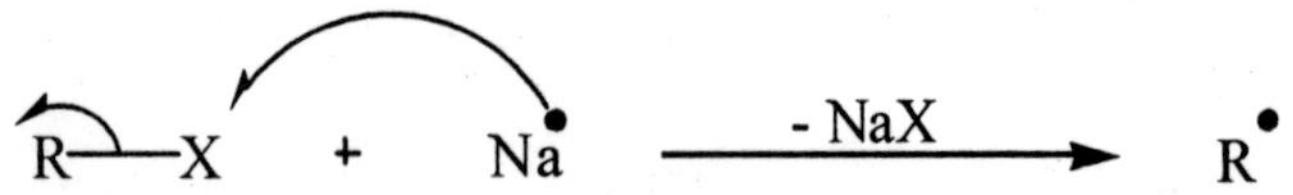

2. Coupling of free radicals takes place to give the product.

$$R^{\bullet} + R^{\bullet} \longrightarrow R—R$$

Small amounts of side products are also obtained by self coupling of phenyl and alkyl residues.

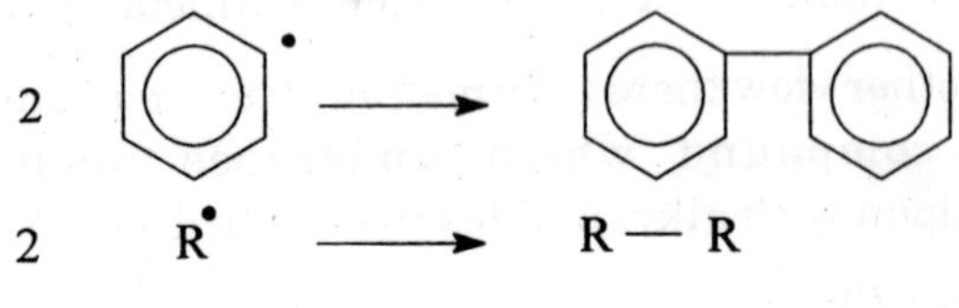

When a mixture of two different alkyl halides is used, a mixture of alkanes is obtained which is difficult to separate.

In another view, there is formation of intermediate organo sodium compound which undergoes nucleophilic substitution with alkyl halides to form the hydrocarbon.

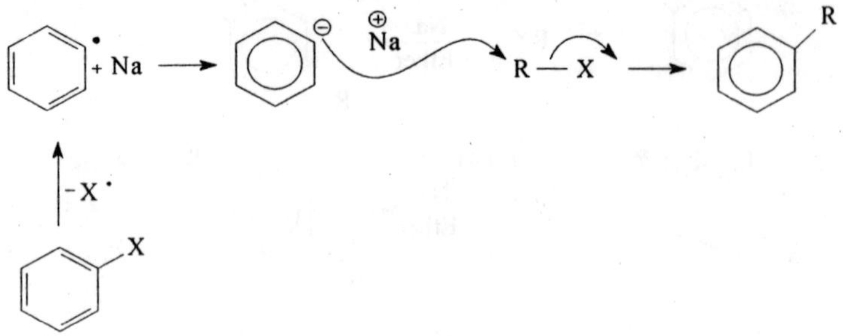

6.9. Zimmermann reaction

The reaction takes place between aromatic nitro compound and methylene ketones in the presence of alkali. The reaction is specially applicable to steroids. When applied to 17- oxosteroides the coloured compounds formed can be used for the quantitative determination of 17-oxosteroids.[29]

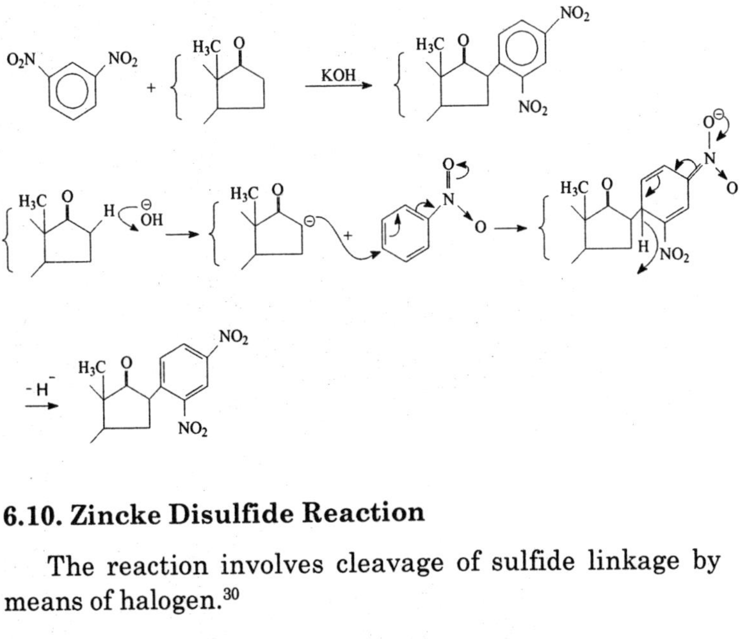

6.10. Zincke Disulfide Reaction

The reaction involves cleavage of sulfide linkage by means of halogen.[30]

$$Ph-S-S-Ph + X-X \longrightarrow 2\,Ph\,S\,X$$

$$Ph-S-H + X-X \longrightarrow Ph\,S\,X + H\,X$$

$$Ph-S-CH_2-Ar + X-X \longrightarrow Ph\,S\,X + Ar-CH\,X_2 + HX$$

6.11. Zincke Nitration

The reaction involves displacement of ortho substituted halogen atoms in aromatic compounds **15** by a nitro group upon treatment with nitrous acid. Substitution by nitrite group takes place when reaction is carried out in acetic acid.[31]

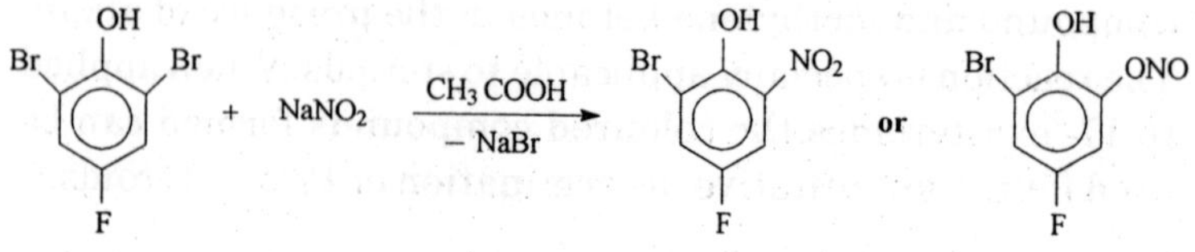

15

6.12. Baker Venkataraman Rearrangement

Diketones **17** are important intermediates in the synthesis of chromones, flavones and are synthesized by base catalysed rearrangement of ortho-acyl ketones **16**[36].

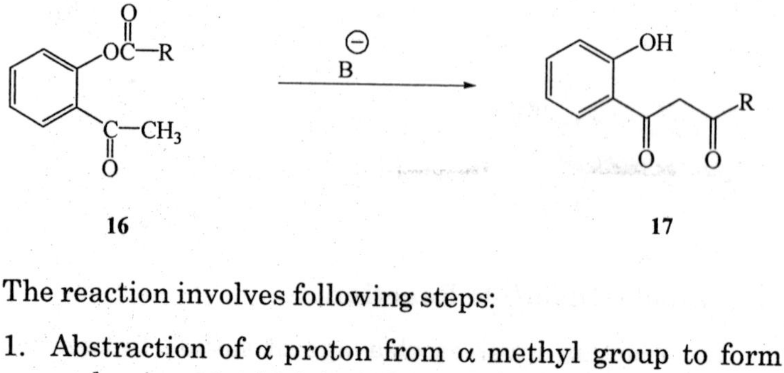

16 17

The reaction involves following steps:

1. Abstraction of α proton from α methyl group to form carbanion **18** which is stabilized by resonance.

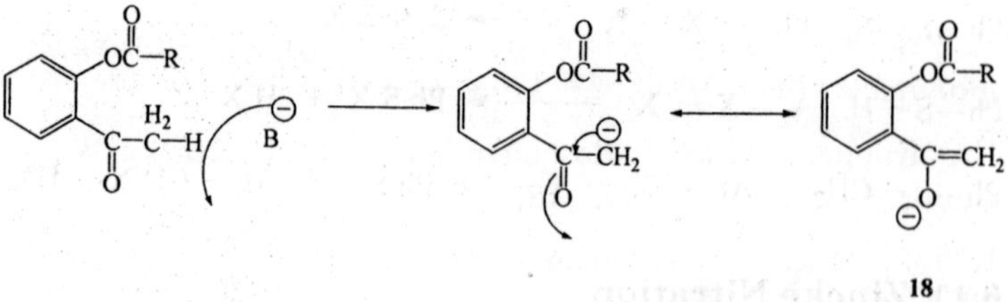

18

2. The carbonion **18** undergoes cyclization followed by ring opening of pyrone ring to form β–diketone **17**.

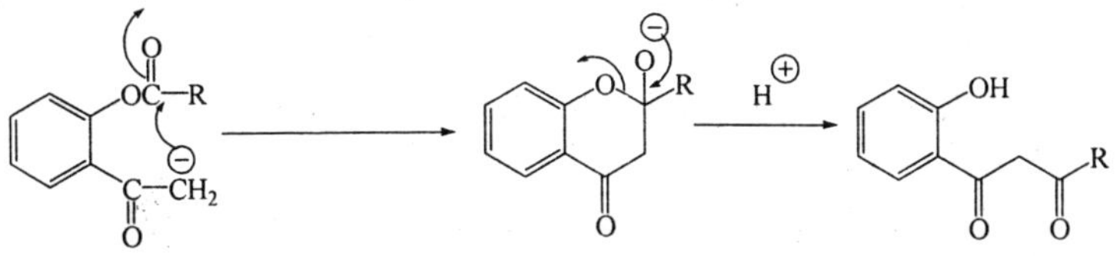

18 17

REFERENCES

1. A. Baeyer, V. Drewon, *Ber.*, 1882, **15**, 2856.
2. B.C. O'Reilly and K.S. Atwal, *Heterocycles*, 1987, **26,** 1185.
3. K.S. Atwal, B.C. O'Reilly, J.Z. Gougoutas and M.F. Malley, *Heterocycles*, 1987, **26**, 1189.
4. K.S. Atwal, G.C. Rovnyak, B.C. O'Reilly and J. Schwartz, *J. Org. Chem.*, 1989, **54**, 5898.
5. H. Bohme, E. Mundlos, *Chem. Ber.*, 1953, **86**, 1414.
6. F. Seet, J.D. Fissekis, *J. Am. Chem. Soc.*, 1973, **95**, 8741.
7. J.D. Fissekis, F. Sweet, *J. Org. Chem.*, 1973, **38**, 1963.
8. A. Takamizawa, K. Hirai, *Chem. Pharm. Bull.*, 1964, **12**, 804.
9. C.O. Kappe, P. Roschger, *J. Heterocycl. Chem.*, 1989, **26**, 55.
10. H. Cho, Y. Takeuchi, M. Ueda, A. Mizuno, *Tetrahedron Lett.*, 1988, **29**, 5405.
11. G. Zigeuner, H. Hamberger, H. Blaschke, H. Sterk, *Monatsch. Chem.*, 1966, **97**, 1408.
12. C.O. Kappe, G. Uray, P. Roschger, W. Linder, C. Kratky, W. Keller, *Tetrahedron*, 1992, **48,** 5473.
13. K. Folkers and T.B. Johnson, *J. Am. Chem. Soc.*, 1934, **56**, 1374.
14. J.J. Boldwin, S.M. Pitzenberger, D.E. McClure, *U.S. Patent*, 1987, 4675, 321; *Chem. Abstr.* 1987, **107**, 242619 d.

15. G. Zigeuner, C. Knopp, H. Blaschke, *Monatsh. Chem.*, 1976, **107**, 587.
16. C.O Kappe, G. Uray,P .Roschger, W. Linder, C.Kratky, W.Keller, *Tetrahedron*, 1992, **48**, 5473.
17. M.K. Jani, N.K. Undavia, P.B. Trivedi, *J. Ind. Chem. Soc.* ,1990, **67**, 847.
18. F.L. Khanna, G. Siliniece, J. Ozols Durbaurs, A. Kimenis, Khim- Farm. *Zh.*, 1978, **12**, 72.
19. G. Zigeuner, H. Hamberger, H. Blaschke, H. Sterk, *Monatsch. Chem.*, 1966, **97**, 1408.
20. C. O. Kappe, *Liebigs Ann. Chem.*, 1990, 505.
21. J. Ashby, D. Griffiths., *J. Chem. Soc., Perkin Trans*, 1975, **1,** 657.
22. T. George, R. Tahilramani, D.V. Mehta *Synthesis,* 1975, 405.
23. G. Zigeuner, C. Knopp, *Monatsh. Chem.*, 1970, **101,** 1541.
24. H.v Pechmann and C. Dutsberg, *Ber.* , 1883**, 16,** 2119.
25. E. Petschek and H. Simmonis, *Ber.*, **46**, 2014, 1913.
26. N. Miyaura and A. Suzuki, *Chem. Rev.*, 1995, **95**, 2457.
27. A. Furstner and G. Seidel, *Synlett,* 1998, **161**.
28. P.S. Wharton and D.H. Bohlen, *J. Org. Chem.* ,1961, **26**, 3615.
29. W. Limmermann, *Z. Physiol. Chem.* ,1935, **233**, 257.
30. T. Zincke, *Ber.* ,1911, **44,** 769.
31. T. Zincke, *J. Prakt,. Chem.*, 1900, **61**, 561.
32. H. Kolancilar, *T.U.J.Sci.Res.B.*, 2002, **3(1)**, 7.
33. G. Smitha and C. S. Reddy, *Synth. Commun.* , 2004, **34(21)**, 3997.
34. E. V. Stoyanov and J. Mezger, *Molecules*, 2005, **10,** 762.
35. S. Kumar, A. Saini and J.S. Sandhu, *Arkivoc*, 2007, **xv**, 18.
36. W. Baker, *J. Chem. Soc.* , 1933, 1381; H. S. Mahal and K. Venkataraman, *ibid*, 1934, 1767.

7

Synthesis of Carboxylic Acids

7.1. Arndt-Eistert synthesis

The reaction consists of following steps:

1. Formation of acid chloride **2** by treatment of carboxylic acid **1** with thionyl chloride or phosphorous pentachloride.

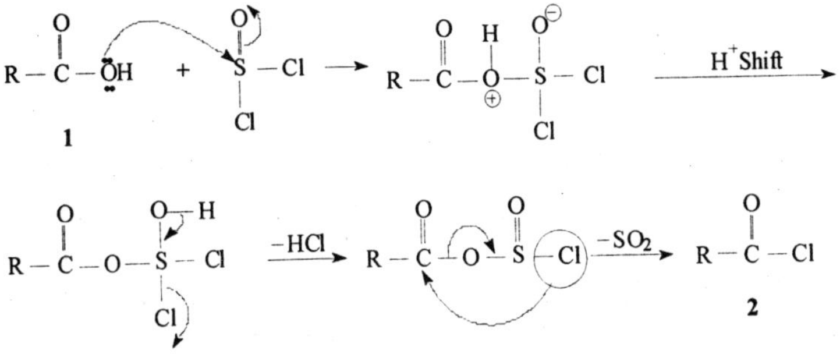

2. Conversion of acid chloride **2** to diazoketone **3** by treatment with diazomethane.

3. Diazoketone **3** undergoes Wolff rearrangement to form intermediate ketene **4** which can be converted to carboxylic acid **5** upon treatment with colloidal silver,

ester **6** upon treatment with alcohol, amides **7, 8** when treated with ammonia or amines.

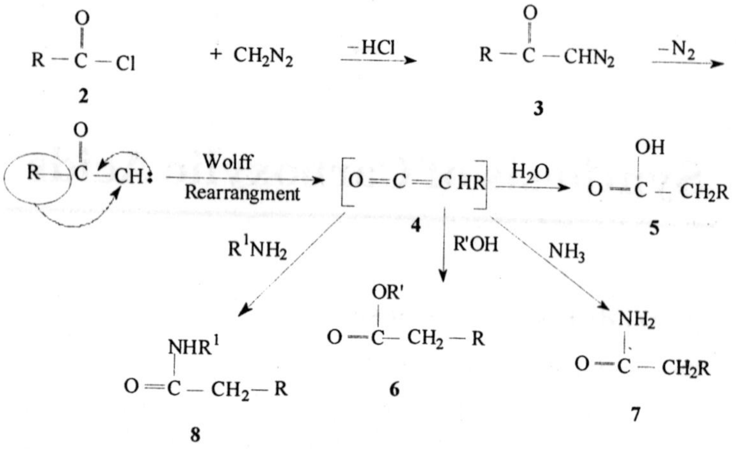

7.2. Barbier-Wieland degradation

The reaction is used to synthesize lower homologue of carboxylic acid by carrying out stepwise degradation of carboxylic acid.[1] The reaction has been generally, observed in steroid side chain.

The substrate carboxylic acid **9** is first converted into its ester **10** which upon treatment with Grignard reagent forms tertiary alcohol **11**. Dehydration of tertiary alcohol in the presence of acetic anhydride gives alkene **12**. The oxidation of **12** with chromic acid gives lower homologue of carboxylic acid **13**.

$$R-CH_2COOH \xrightarrow[\text{degradation}]{\text{Barbier Wieland}} RCOOH$$

The sequence of reaction is depicted below:

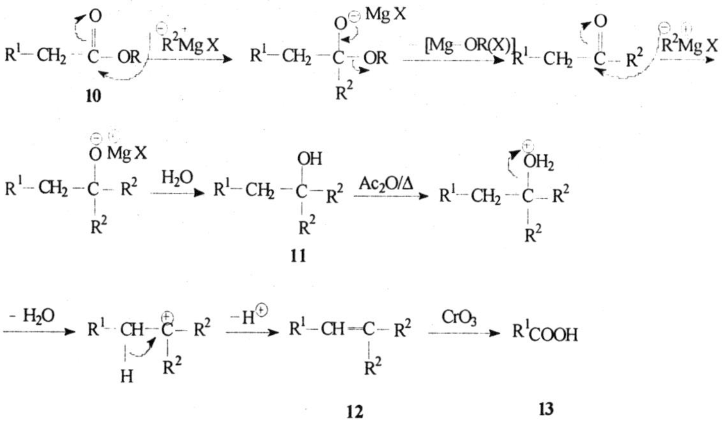

7.3. Hell-Volhard Zelinsky reaction

The reaction consists in treating a carboxylic acid containing α H with halogen (chlorine, bromine) in the presence of phosphorus. The product is α- halocarboxylic acid **14** and the reaction is called as Hell-Volhard Zelinsky reaction.[2]

$$R-CH_2-\overset{\overset{\displaystyle O}{\|}}{C}-OH \xrightarrow{X_2 + P} RCHX-\overset{\overset{\displaystyle O}{\|}}{C}-OH$$

14

Following steps are involved in the synthesis of α-halocarboxylic acid **14**.

1. Halogen and phosphorous form phosphorous trihalide and acts as a catalyst for the conversion of carboxylic acid into acid nalide.

$$2P + 3X_2 \longrightarrow 2PX_3$$

$$3R-CH_2-\overset{\overset{\displaystyle O}{\|}}{C}-OH + 2PX_3 \longrightarrow 3R-CH_2-\overset{\overset{\displaystyle O}{\|}}{C}-X + H_3PO_3$$

2. Enolization of acid halide in acidic medium takes place to form **15** which reacts with halogen to give **16**.

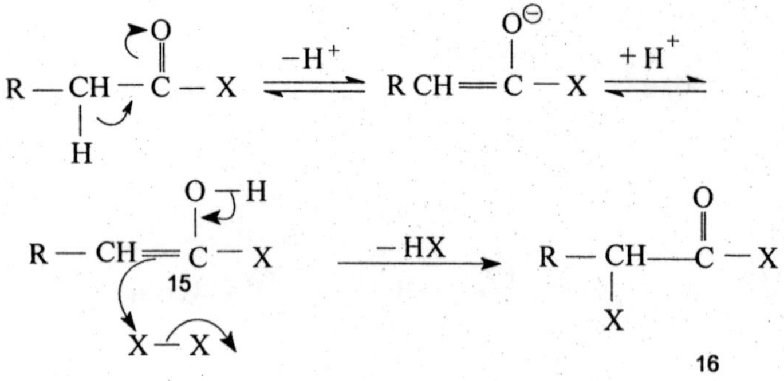

3. The formation of α-halo acid **14** takes place by reaction of α-halo acid halide **17** with carboxylic acid or water.

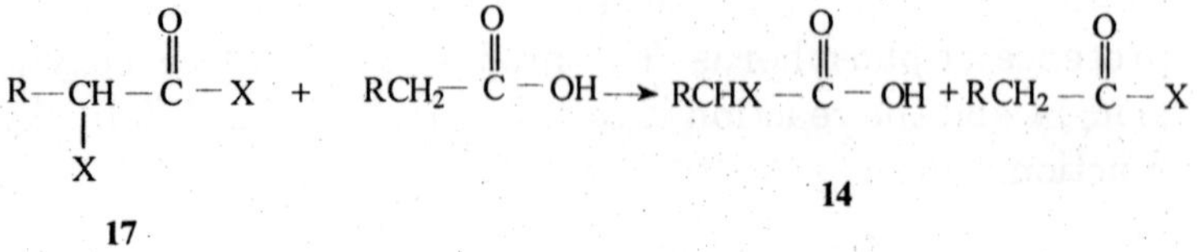

The reaction in general is, thus, used for the synthesis of α-halo carboxylic acid e.g.

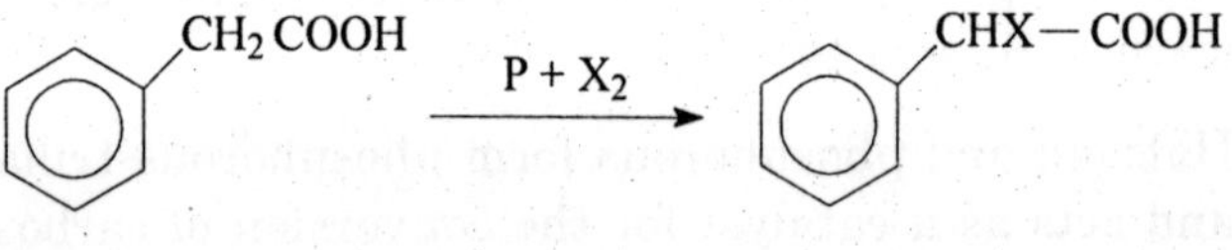

7.4. Perkin reaction

The reaction is used for the synthesis of aromatic α,β unsaturated carboxylic acid. The product **18** is obtained when an aromatic aldehyde **15** is treated with an aliphatic acid anhydride **16** in the presence of sodium or potassium salt of the acid corresponding to anhydride **17**.[3]

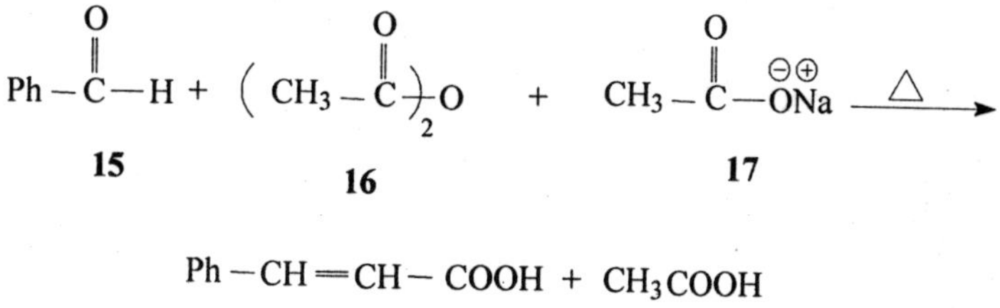

15 **16** **17** **18**

The mechanism of the reaction involves following steps.

1) The carbanion **19** is obtained by removal of α-hydrogen of acetic anhydride **16** by the base.

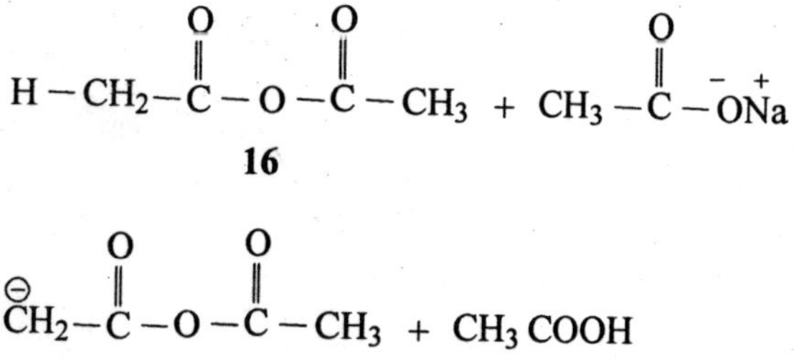

16 **19**

2) Carbanion **19** attacks carbonyl carbon of aldehyde **15** to form **20**.

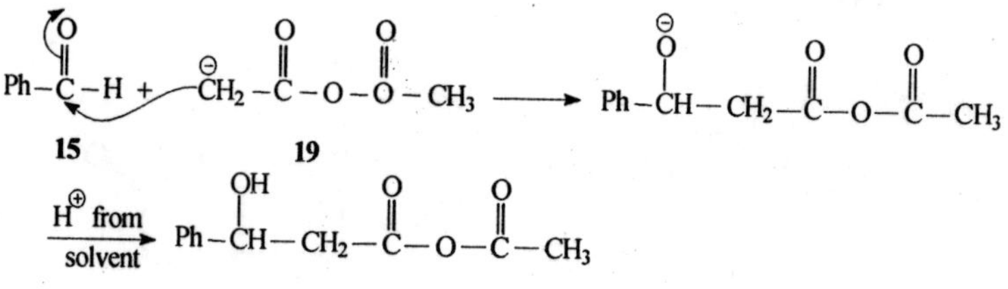

15 **19** **20**

3) Removal of α H by base to form carbanion **21**, followed by the removal of hydroxyl group to give unsaturated anhydride **22**.

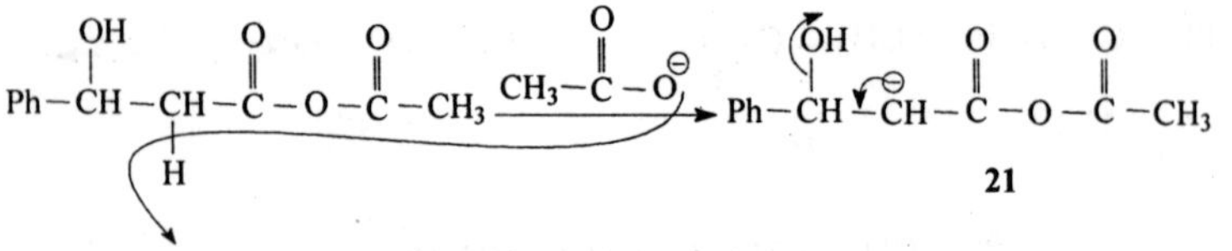

21

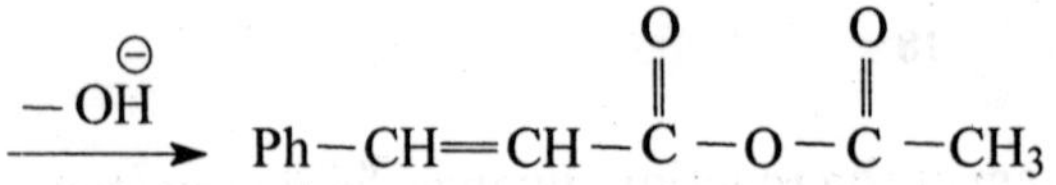

22

4) Hydrolysis of anhydride **22** gives α, β- unsaturated carboxylic acid **23**.

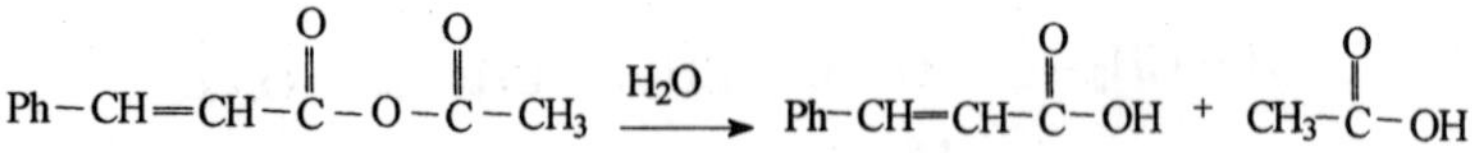

22 **23**

Synthetic Application

Perkin Reaction is used for the synthesis of coumarins **24** from salicylaldehyde, acetic anhydride in the presence of sodium acetate.[4]

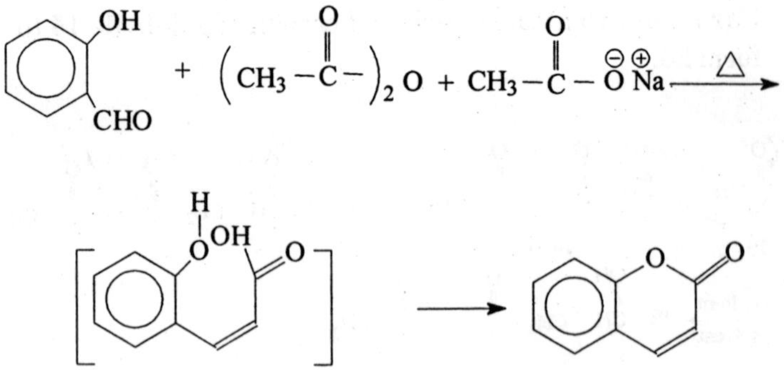

24

REFERENCES

1. H . W ieland, *Ber.*, 1912, 484, P. Barbier and R. Locquin, *Compt. Rend.*, 1913, **156**, 1443.

2. C. Hell, *Ber.*,1881, **14**, 891; J. Volhard, *Ann.*, 1887, **242**, 141; N. Zelinsky, *Ber* ., 1887, **20**, 2026.

3. W.H. Perkin, *J. Chem. Sec.*, 1868, **21 (58),** 181; J. R. Johnson, *Org. Reaction*, 1942, **1**, 210.

4. W. H. Perkin, *J. Chem. Sec.* ,1868, **21 (58),** 181. J.R. Johnson, *Org. Reactions*, 1942, **1**.

8

Synthesis of Esters

8.1. Baeyer-Villiger Oxidation

The oxidation of ketones to esters **1** / lactones **2** with per acid is known as Baeyer- Villiger oxidation.[1]

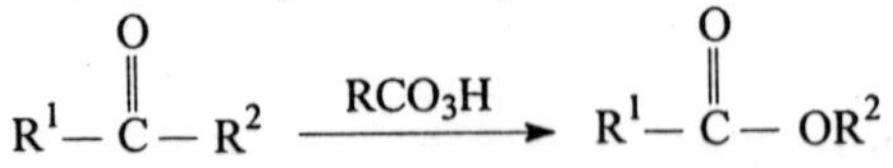

1

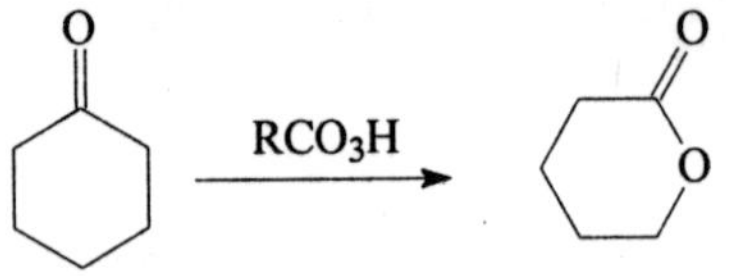

2

The reaction involves following steps.

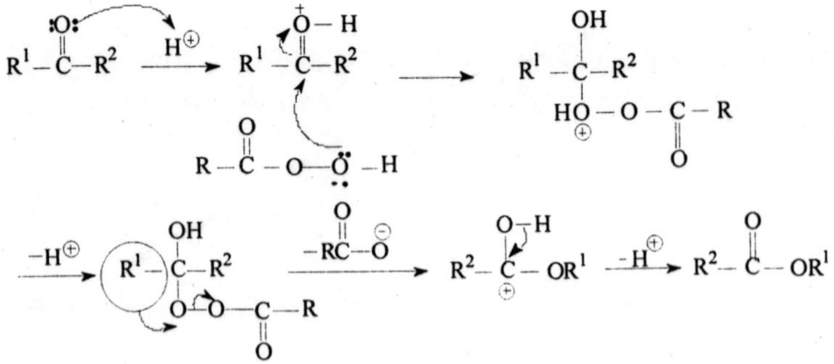

8.2. Darzen's condensation

The condensation of aldehydes or ketones **3** with α-haloesters **4** in the presence of a base (preferably a hindered base) to form epoxyesters **5** (glycidic esters) is known as Darzen reaction.[2]

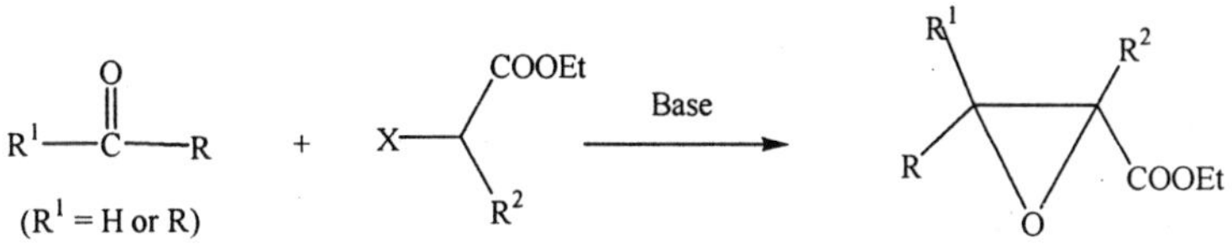

3 4 5

These epoxy esters are thermally unstable and undergo decarboxylation upon heating to form aldehydes or ketones. The mechanism of the reaction involves formation of carbanion **6** of α halo esters **4** upon treatment with base. The formation of carbanion **6** is feasible because of its stabilization by resonance.

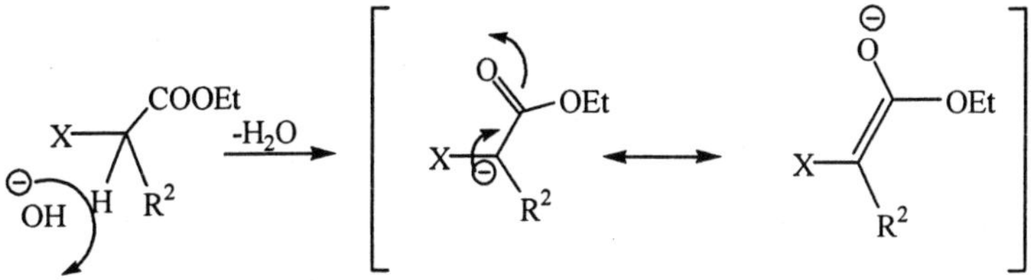

4 6

The carbanion **6** attacks carbonyl carbon of aldehyde or ketone to form **7** which expels halide group through nucleophilic substitution to form the product **5**.

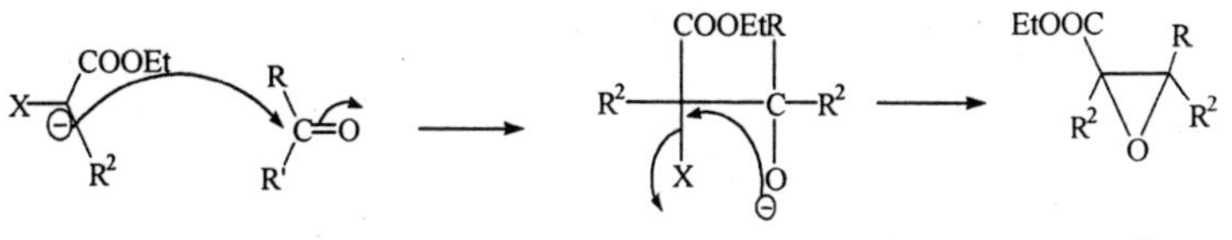

6 3 7 5

Glycidic ester **5** upon heating with aquous base or acid undergoes decarboxylation and forms aldehyde or ketone.

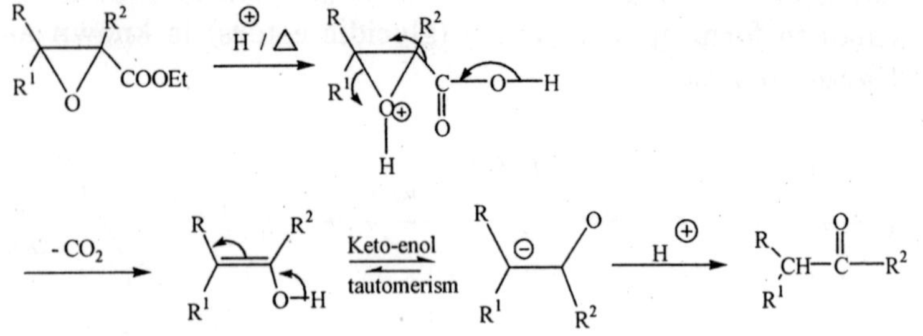

8.3. Reformatsky reaction

The reaction is used for the synthesis of β-hydroxyester from α-haloester and carbonyl compound (aldehydes / ketones). Thus, when α-haloester **9** is treated with carbonyl compound **8** in the presence of zinc, β-hydroxyester **10** is obtained.

$$\underset{\mathbf{8}}{-\overset{\overset{\displaystyle O}{\|}}{C}-} + \underset{\mathbf{9}}{Br-CH_2-COOR} \xrightarrow[\text{(ii) } H_2O]{\text{(i) Zn / Ether}} \underset{\mathbf{10}}{-\overset{\overset{\displaystyle OH}{|}}{\underset{|}{C}}-CH_2-COOR}$$

The mechanism of the reaction involves following steps:

1. Formation of organozinc compound **11** by interaction of α-haloester **9** with zinc in a non polar solvent such as ether or benzene.

$$Br-CH_2COOR + Zn \xrightarrow{\text{Ether}} \underset{\mathbf{11}}{\overset{\oplus}{Br}\,\overset{\ominus}{Zn}\,CH_2\,COOR}$$

2 Organozinc compound **11** reacts with carbonyl compound

(aldehyde / ketone) **8** to form addition product which upon hydrolysis forms β hydroxy compound **12**.

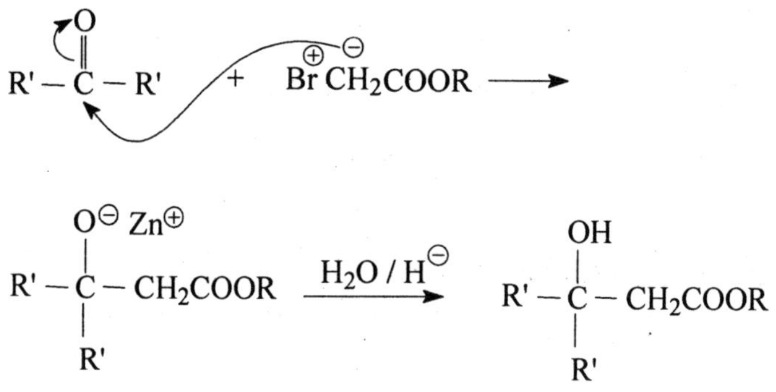

12

REFERENCES

1. A. Baeyer, V. Villiger, *Ber.*, 1899, **32**, 3625; *ibid.*, 1900, **33**, 855.

2. G. Darzen, *Compt. Rend.*, **139**, 1214, 1904.

9

Synthesis of Olefins / Alkynes

9.1. Bamford Stevens reaction

The reaction is used for the formation of olefins by base promoted decomposition of p-toluenesulfonyl hydrazones of carbonyl compounds (aldehydes and ketones).[1]

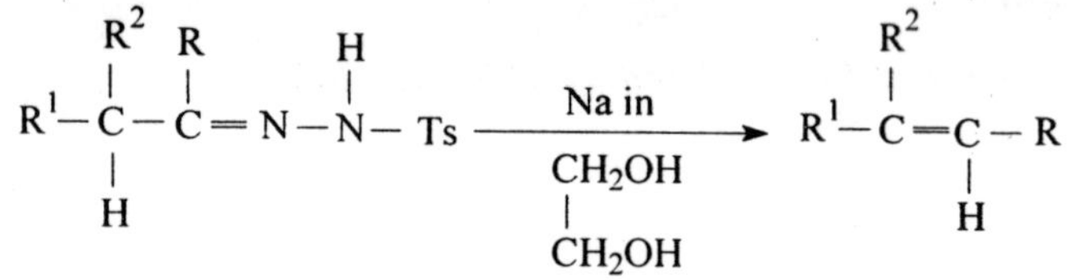

Two kinds of mechanisms have been proposed:

(a) *Anionic mechanism*; In this type of mechanism a strong base such as organolithium compound ($R^{-}Li^{+}$) is used and least substituted alkenes are formed (also known as Shapiro reaction).[2]

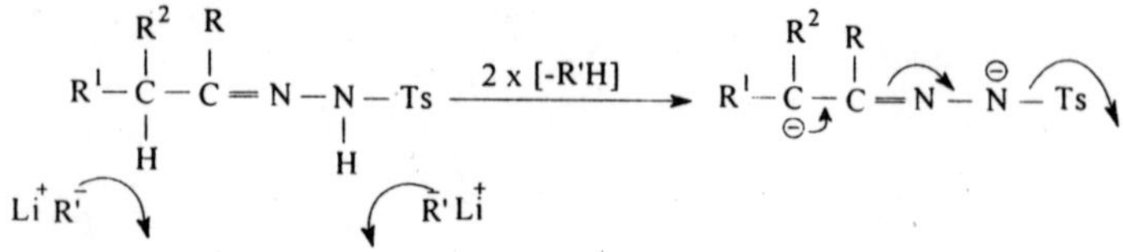

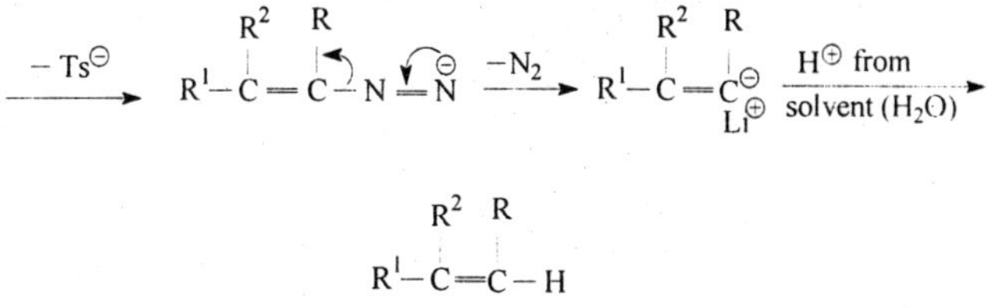

(b) The reaction mechanism involving carbene as intermediate involves following steps:

1. Tosylhydrazone of carbonyl compound upon treatment with base forms diazo compound **1** with subsequent expulsion of tosyl group.
2. Elimination of nitrogen from diazo compound **1** forms carbene **2** which undergoes 1,2 hydride shift to form alkene **3**.

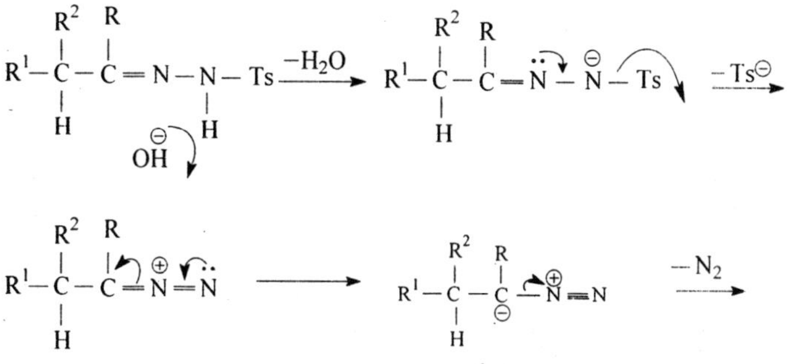

1

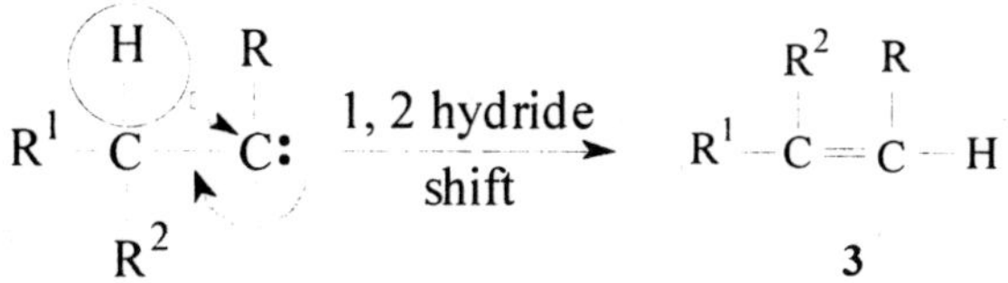

. 2 3

9.2. Birch reduction

The reaction is used for reduction of aromatic ring to unconjugated dihydroderivative in the presence of alkali metal(Li, Na, K etc.) in liquid ammonia or amines.[3] The reaction is conducted in protic solvent such as alcohols.

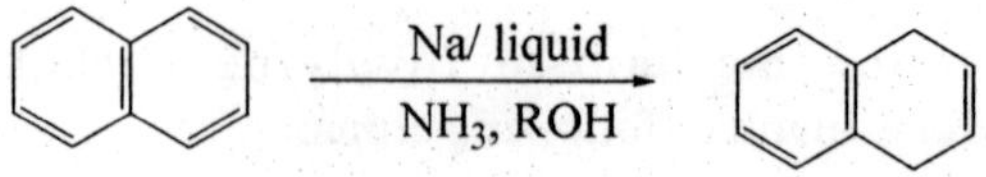

1,4-Dihydronaphthalene

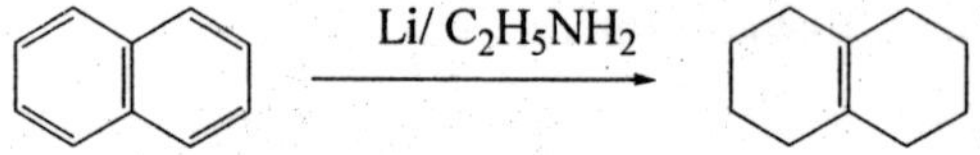

9, 10-Octalin

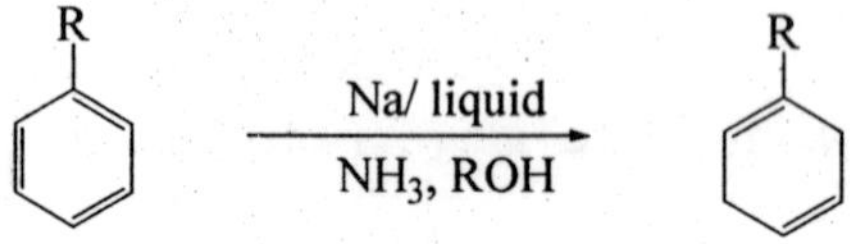

R may be electron releasing or electron withdrawing group. Generally aromatics rings, conjugated alkenes, and alkynes are reduced but alkanes with isolated double bond are not affected. Alkynes **4** are reduced to trans olefins **5**.

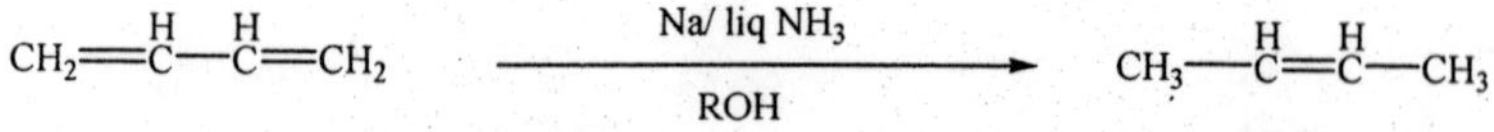

4

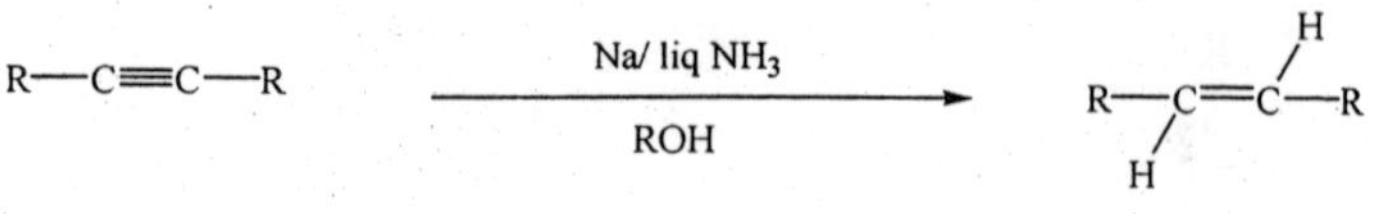

5

The mechanism involves following steps

i) Metals deliver their electrons present in outermost shell to Π electron system to form radical anion **6**.

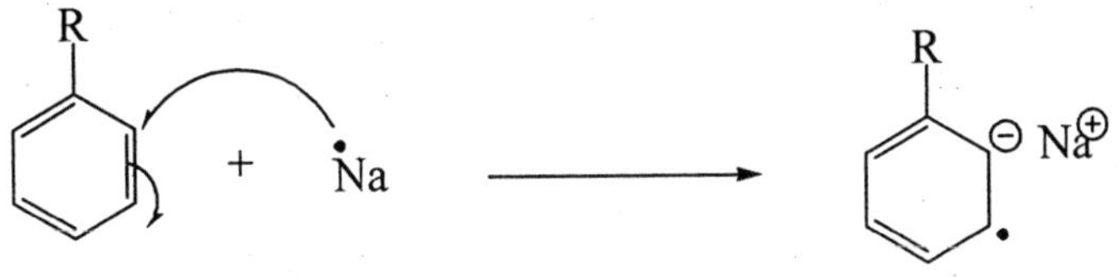

6

ii) If proton source is available, the radical anion **6** gets protonated to give **7**.

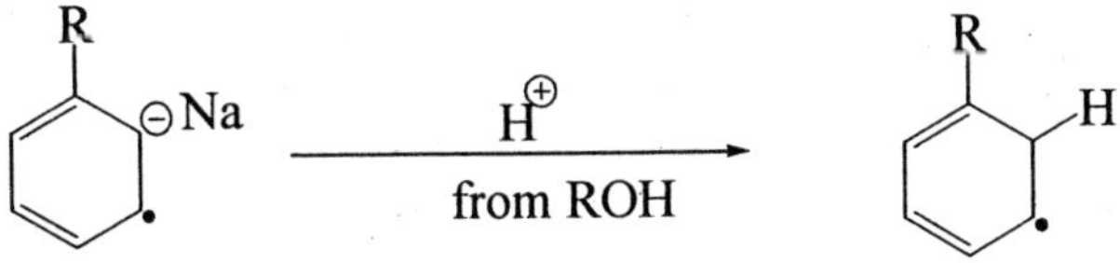

7

iv) The compound **7** is further reduced by metal to give anion **8** which takes proton from solvent to give **9**.

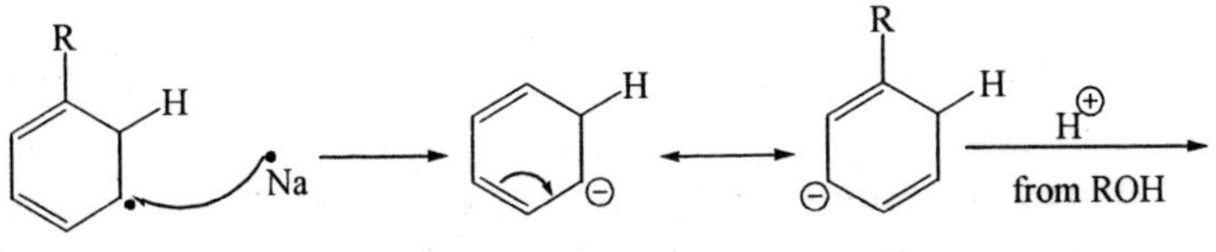

7 **8**

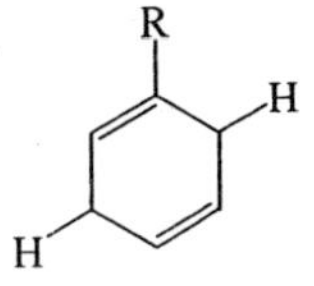

9

The reduction of arylethers followed by hydrolysis serves as method for synthesizing cyclohexanones.

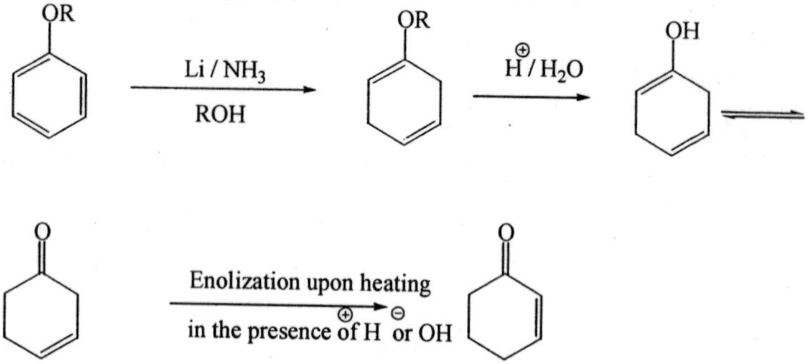

Synthetic applications of Birch reduction

1. The Birch reduction of 2-methoxy-1-naphthoic acids **10** with 2.4 equivalents of sodium in liquid ammonia leads to the formation of 1,4-dihydro compounds **11**.[4]

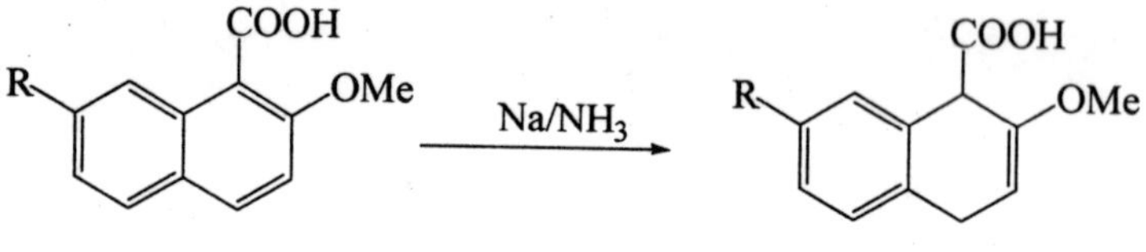

10 **11**

2. The Birch reduction has been also used in the total synthesis of natural products such as stemphol **12**, altenusin **13**, alternariol **14**, mellein **15**, 6-methoxymellein **16**, curvularin **17**, naturally occurring phytotoxic phthalides **18** and xanthone morellin **19**.[5]

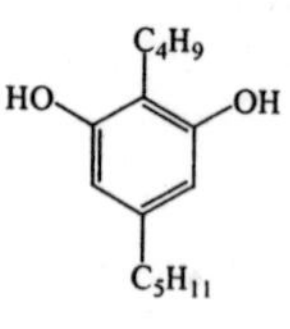

12

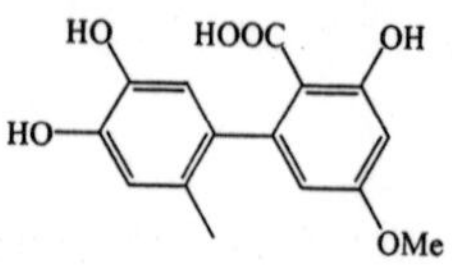

13

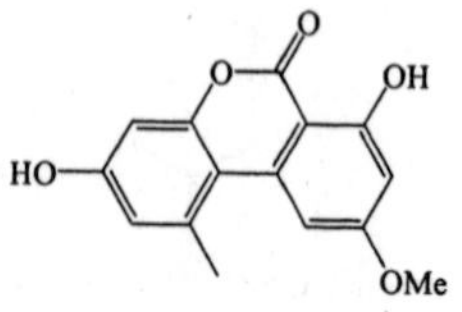

14

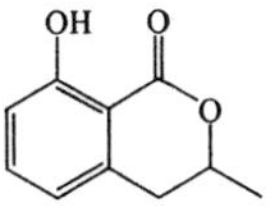

15

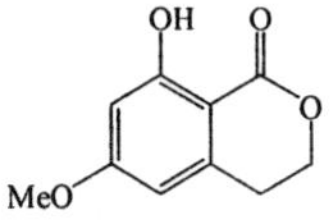

16

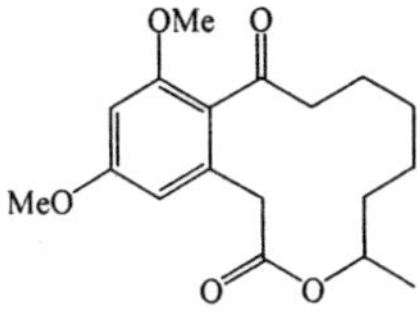

17

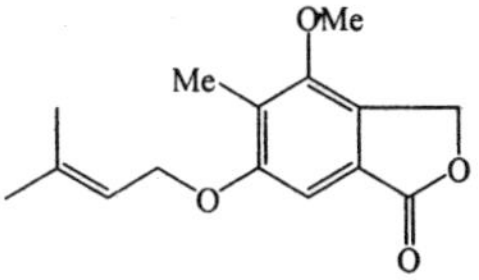

18

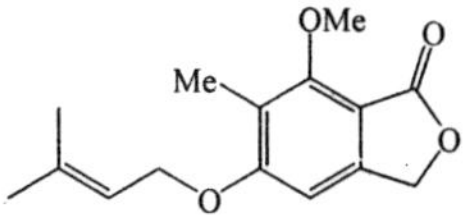

19

9.3. Glaser coupling

The reaction involves homocoupling of terminal alkynes. The reaction is catalysed by cuprous salts in the presence of ammonium chloride and an oxidant.[6,7]

$$R{-}C{\equiv}C{-}H \xrightarrow[NH_4Cl]{CuCl/O_2} R{-}C{\equiv}C{-}C{\equiv}C{-}R$$

Hetercoupling is also possible if the reaction occurs between terminal alkynes and halo alkynes(Cadiot-Chodkiewicz coupling).

$$R{-}C{\equiv}C{-}H + Br{-}C{\equiv}C{-}R' \xrightarrow{CuCl/RNH_2} R{-}C{\equiv}C{-}C{\equiv}C{-}R'$$

The heterocoupling is also carried out in the presence of cupric salts in pyridine (Eglinton reaction). Glacer coupling is also reported in super critical carbon dioxide($ScCO_2$) in the presence of solid base such as sodium acetate(NaOAc).[8]

9.4. Heck reaction

The coupling of an organic halide which does not possess

an sp^3 hybridized carbon and sp^3 hybridized β hydrogen with an alkene in the presence of palladium is called as Heck reaction.[9-11] The reaction is highly stereospecific.

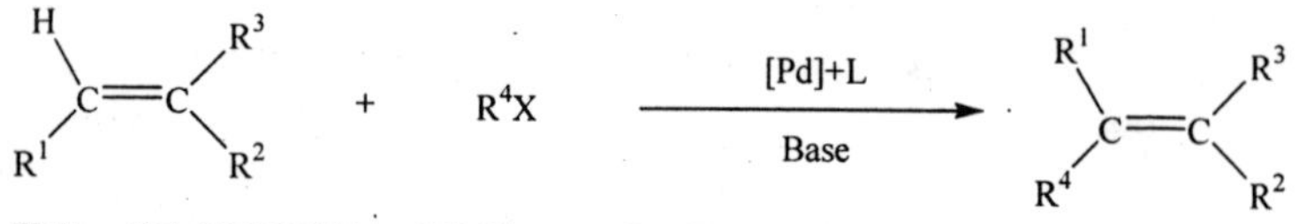

[Pd] = Pd $(OCOCH_3)_2$, $PdCl_2$, etc.,L= PAr_3 etc.

Base= $N(Et)_3$, K_2CO_3, NaOAc.

R^4= Aryl, alkenyl, benzyl

X= I, Br, OSO_2CF_3

Thus, stilbene **24** has been synthesized by coupling of styrene **22** with bromobenzene **23** in the presence of palladium acetate.[12]

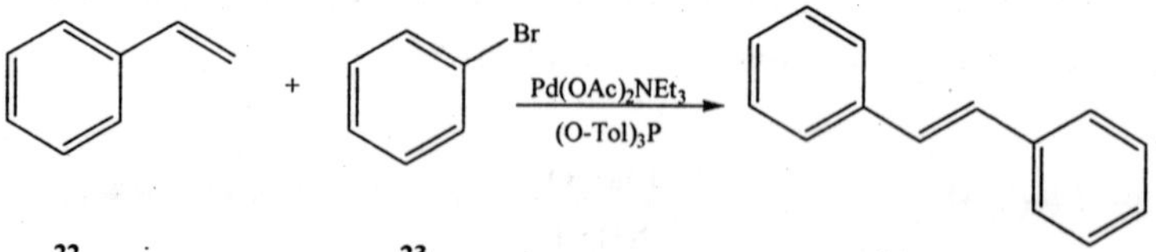

22 **23** **24**

Cyclization of halide **25** to **26** takes place under Heck reaction conditions.[13]

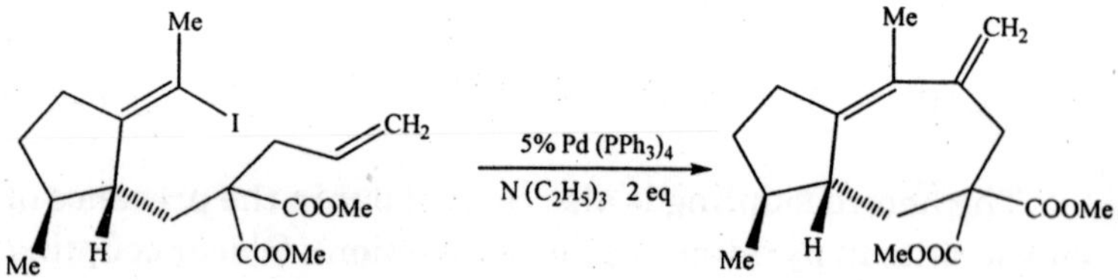

25 **26**

9.5. Hofmann degradation reaction

The reaction is also known as exhaustive methylation.

The reaction involves formation of an olefin **27** and a tertiary amine **28** by heating at high temperature of quaternary ammonium hydroxide.[14]

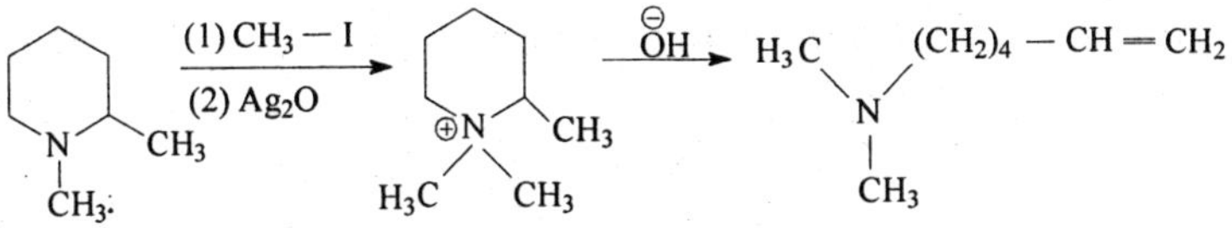

The reaction takes place by abstraction of a β-hydrogen in the form of proton by base followed by formation of π bond and simultaneous expulsion of trialkyl amine. Thus, it is E_2-elimination as formation and breaking of bonds takes place simultaneously. Hofmann elimination products are generally least substituted alkenes (governed by Hofmann rule).

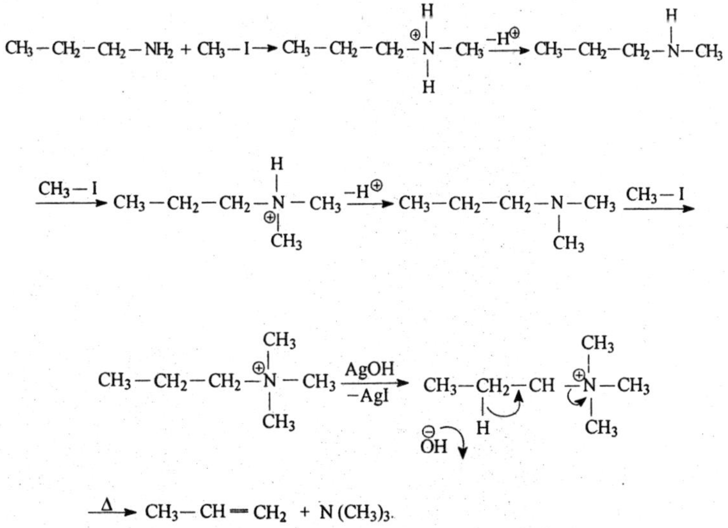

27 **28**

Hofmann degradation method is generally useful for the structure determination of natural products such as alkaloids.

9.6. Hofmann Elimination

The reaction involves exhaustive alkylation of amines with alkyl iodides to give quaternary ammonium salts which when treated with silveroxide in the presence of a base undergoes a elimination to give least substituted alkenes.[15]

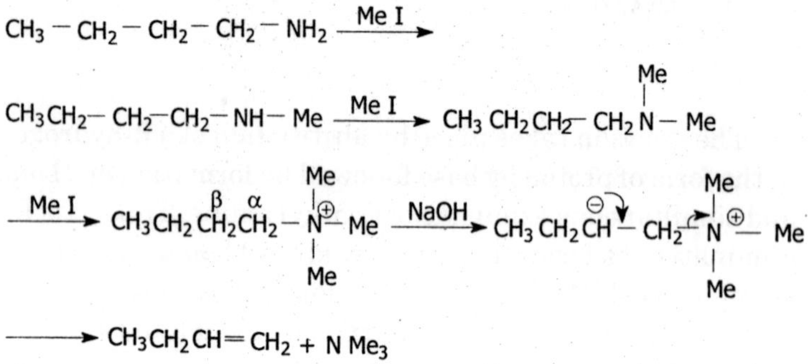

The cyclic alkene **30** has been obtained in quantitative yield from tertiary amine **29** through Hofmann elimination.[16]

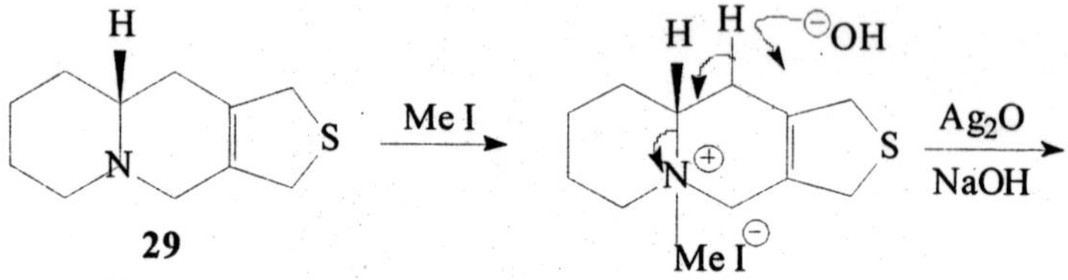

29

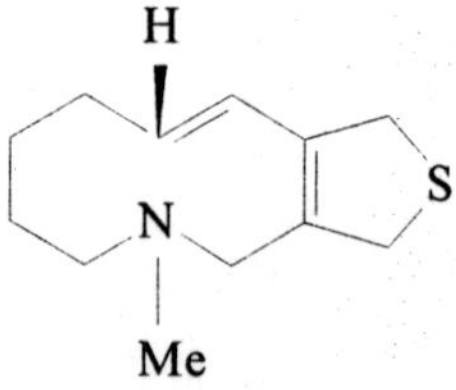

30

9.7. Horner-Wadsworth- Emmons Olefination

In this reaction an aldehyde is converted to an alkene by using phosphate derivative.[17,18]

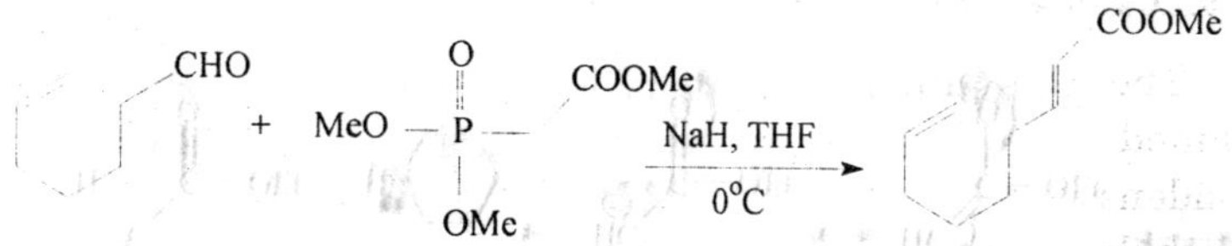

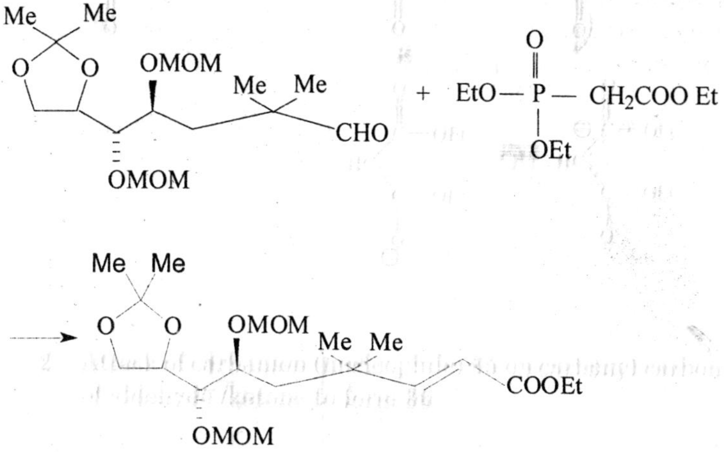

9.8. Keck allylation

The reaction involves allylation of aldehyde **31** by employing allyltributyltin in BF_3-Et_2O at low temperature.[19,20]

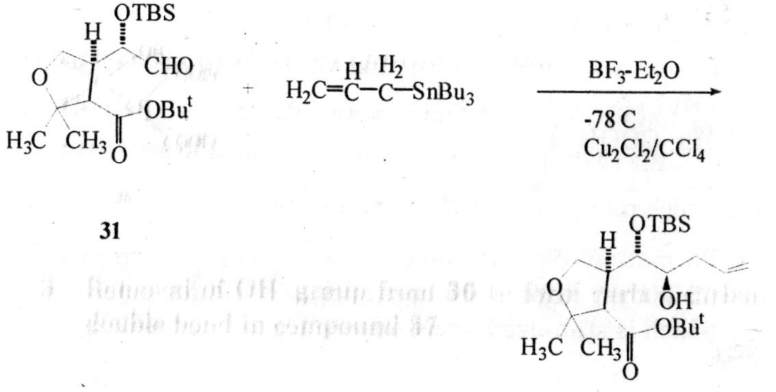

31

95%

9.9. Knoevenagel condensation

The Knoevenagel condensation is an organic reaction named after Emil Knoevenagel. It is a modified aldol condensation.[21] It involves nucleophilic addition of an active methylene compound **32** to an aldehyde or ketone followed by dehydration of the intermediate hydroxyl compound **33** to give an α,β-unsaturated compound **34**. The reaction takes place in the presence of a base.

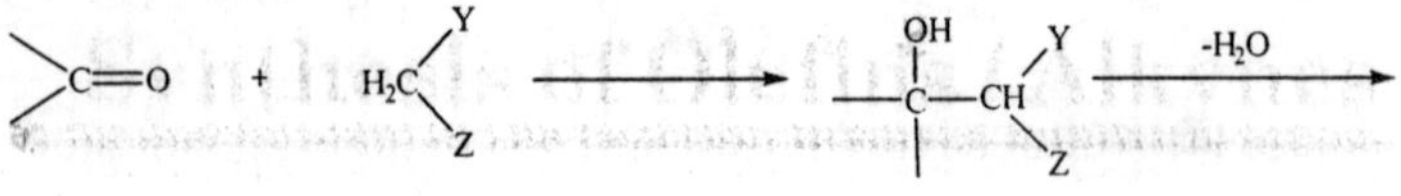

32 **33**

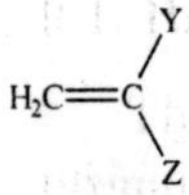

34(a-d)

34a. Y = $COCH_3$, Z = $COOC_2H_5$

34b. Y = Z = $COOC_2H_5$

34c. Y = Z = CN

34d. Y = Z = COOH etc.

In active methylene compound **32** both Y and Z represent powerful electron withdrawing groups such as, CN, NO_2, COOEt, COOH etc.[22]

The mechanism involves following steps:

1. Formation of carbanion of active methylene compound **35** in the presence of a base such as pyridine, piperidine which is stabilized by resonance e.g.

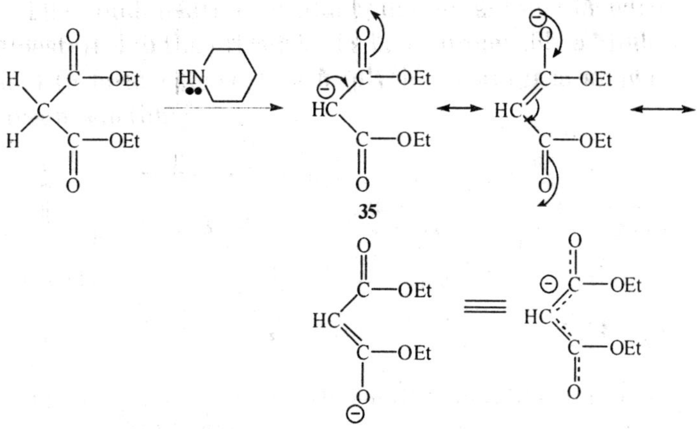

2- Attack of carbanion (nucleophile) **35** on carbonyl carbon of aldehyde /ketone to form **36**.

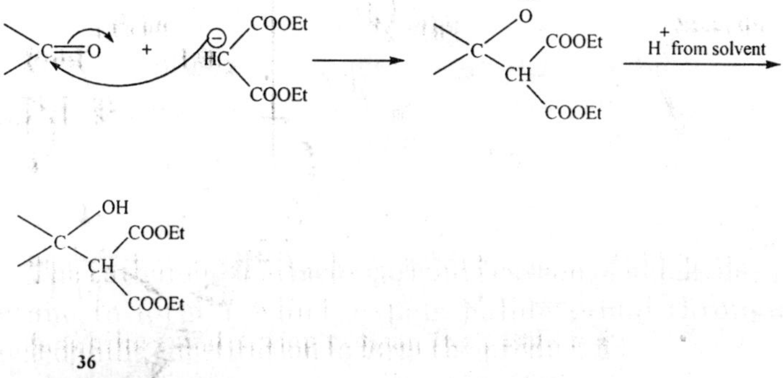

3 Removal of OH^- group from **36** to form carbon-carbon double bond in compound **37**.

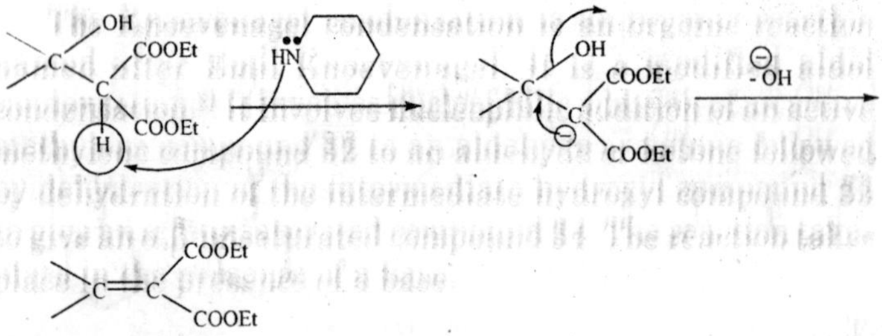

37

Synthetic applications of Knoevenagel Condensation reaction

1. Synthesis of **38** by the reaction of 2-methoxybenzaldehyde with thio barbituric acid in ethanol in the presence of piperidine.[23]

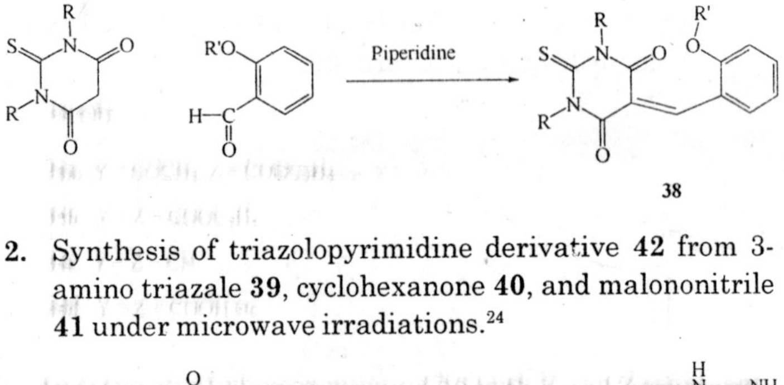

38

2. Synthesis of triazolopyrimidine derivative **42** from 3-amino triazale **39**, cyclohexanone **40**, and malononitrile **41** under microwave irradiations.[24]

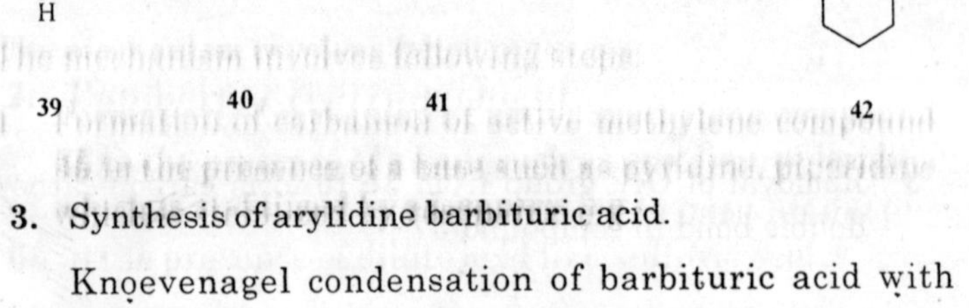

39 40 41 42

3. Synthesis of arylidine barbituric acid.

Knoevenagel condensation of barbituric acid with

aldehydes **43** has been reported to occur both in refluxing aqueous solution as well as under microwave irradiation.[25,26]

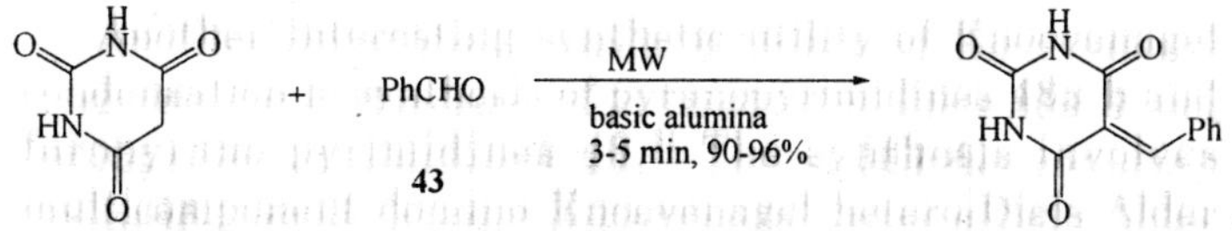

4. *Knoevenagel condensation reaction in microreactors using a new zeolite catalyst:* The CsNaX-NH_2 Catalyst (new basic zeolite), obtained by grafting amino groups into NaX and CsNaX zeolite shows excellent catalytic activity for Knoevenagel condensation .[27] Thus ,higher yield of products are obtained as compared to amino propylated MCM-41 catalyst.

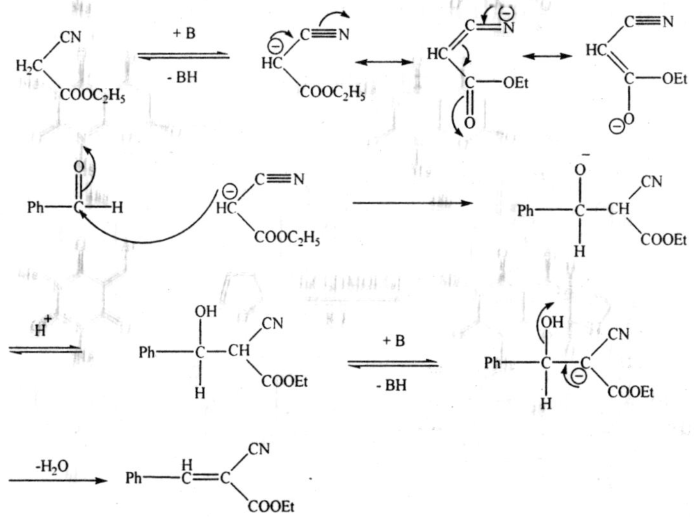

5. *Synthesis of α- cyanoacrytates and α-cyano acrylonitriles.* Knoevenagel condensation gives

α–cyanoacrylates and cyano acrylonitriles **44** from aldehydes and active methylene compound containing CN group in the presence of triphenyl phosphine under conventional as well as microwave irradiation .[28]

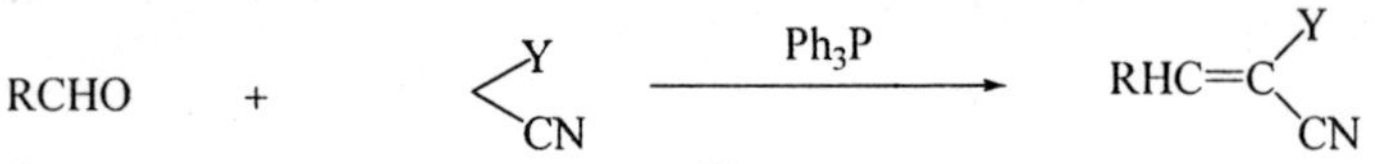

Y= CN, COOEt

44

6. *Synthesis of 5-arylidenyl-2,4-thiazolidinedione under given condition:*

Knoevenagel condensation of various aldehydes and 2,4-thiazolidinedione is carried out using polyethylene glycol-30 under mild conditions. The reaction involves simple work up procedure with better yields.[29]

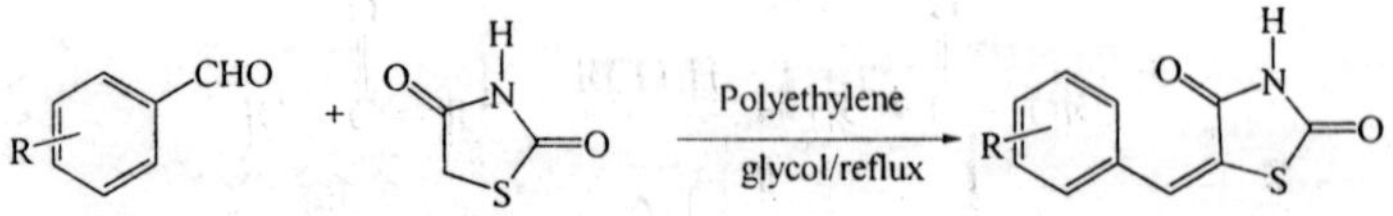

7. *Knoevenagel condensation under solvent free and aquous conditions:-*

In recent years chemists are continuously interested in organic synthesis under green condition. Thus ,Knoevenagel condensation reactions have been also explored under solvent free and aqueous conditions. Thus, authors have reported using ketone **45** instead of aldehyde which is usually used in Knoevenagel condensation under solvent free condition to form alkene **46**.[30]

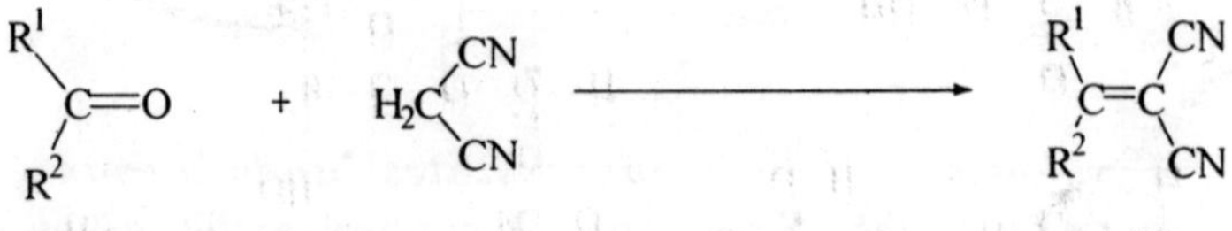

45 **46**

8. *Synthesis of novel pyrano[2,3-d] and furopyrano[2,3-d]pyrimidines*

Another interesting synthetic utility of Knoevenagel condensation is synthesis of pyranopyrimidines **48a-b** and furopyrano pyrimidines **49**.[31] The synthesis involves multicomponent domino Knoevenagel hetero-Diels Alder reaction utilizing the starting materials, an aromatic aldehydes **47**, 1,3-dimethyl barbituric acid and ethyl vinyl ether/2,3-dihydrofuran in the presence of Indium (III) chloride at room temperature.

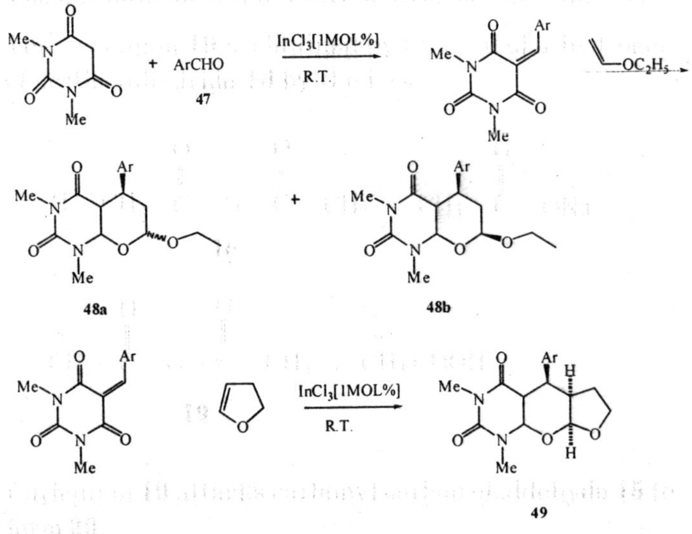

8. *Synthesis of Nitrostyrene*

Nitro styrene **51** are obtained conveniently by Knoevenagel condensation from an aromtic aldehyde **50** and a nitro derivative.[32]

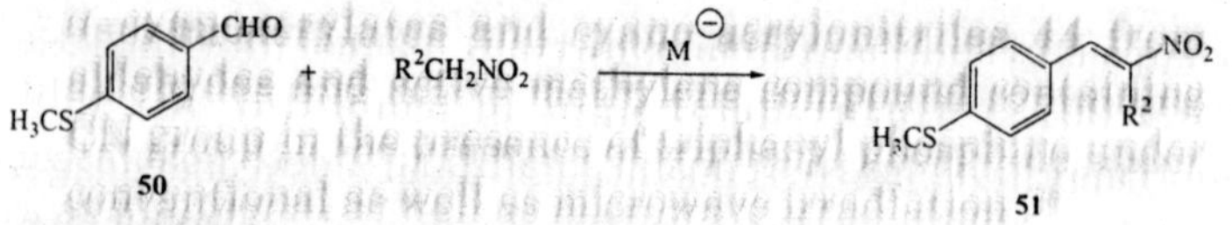

9. *Synthesis of pyrazolinoquinolizine and 1,4-oxazinopyrazoline derivatives*

A variety of fused heterocycles **53** is obtained by the Knoevenagel condensation of 5-tert-amino-3-methyl-1phenyl pyrazole-4-carboxaldehyde **52** with active methylene compounds such as diethylmalonate and cyanoacetamide and subsequent cyclization using anhydrous $ZnCl_2$.[33]

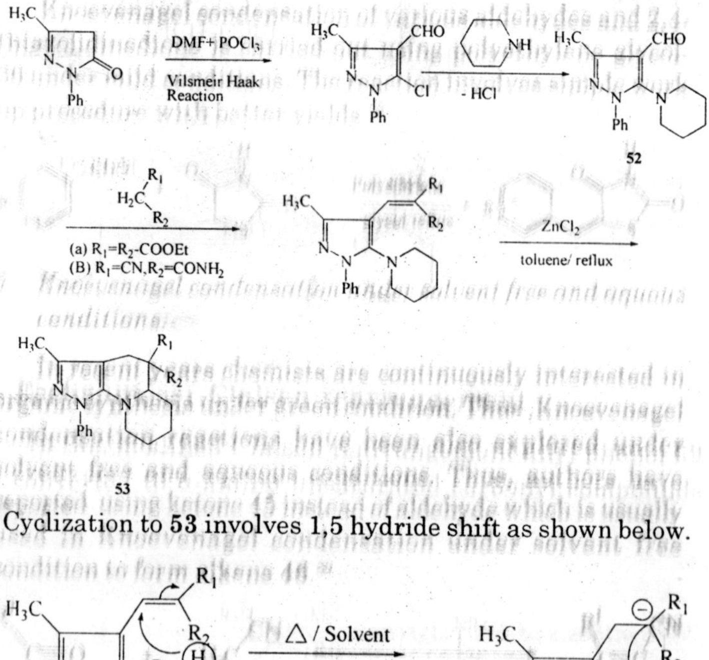

Cyclization to **53** involves 1,5 hydride shift as shown below.

10. *Knoevenagel condensation catalysed by Clay*:

Knoevenagel condensation of Meldrum's acid with aldehydes in the presence of Clay *viz.* Kaolin at room temperature and in the absence of solvent has been also reported. The novelty of the reaction lies in its simplicity and high yields of the pure products.[34]

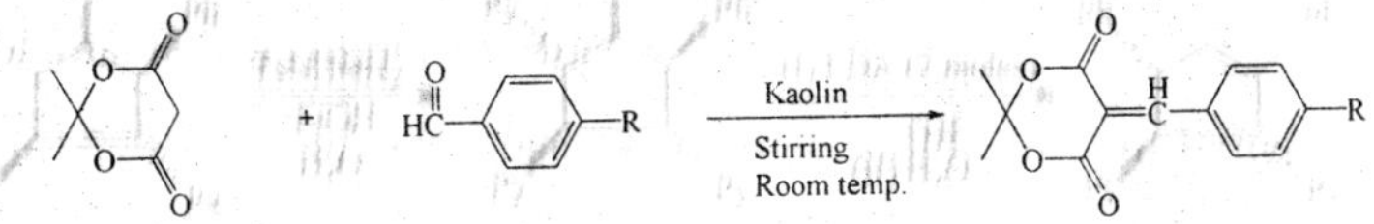

9.10. McMurry Coupling

The coupling of dialdehydes to form a cyclic alkene in the presence of $TiCl_4$ and zinc is called as McMurry coupling.[35,36]

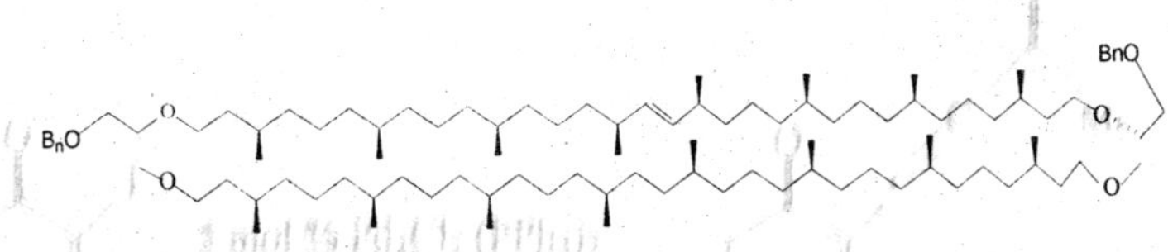

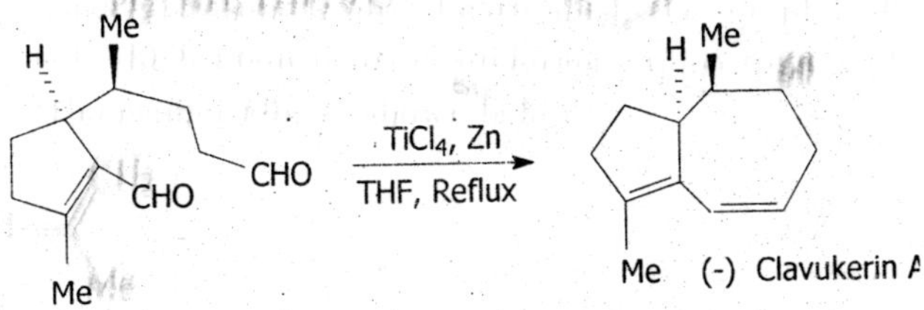

9.11. Ramberg-Backlund reaction

The reactions involves synthesis of alkenes **55** from α-halosulfones **54** in the presence of a strong base.[37]

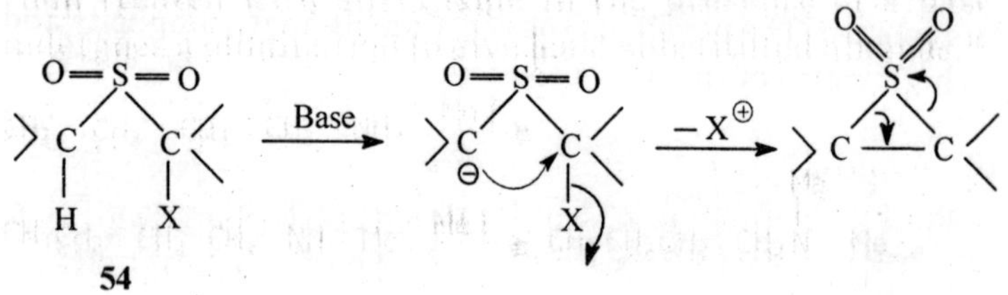

54

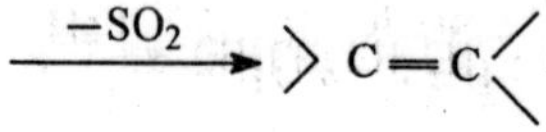

55

The alkene **57** has been synthesized from sulfone derivatives **56** in the presence of morpholine in similar way.[38]

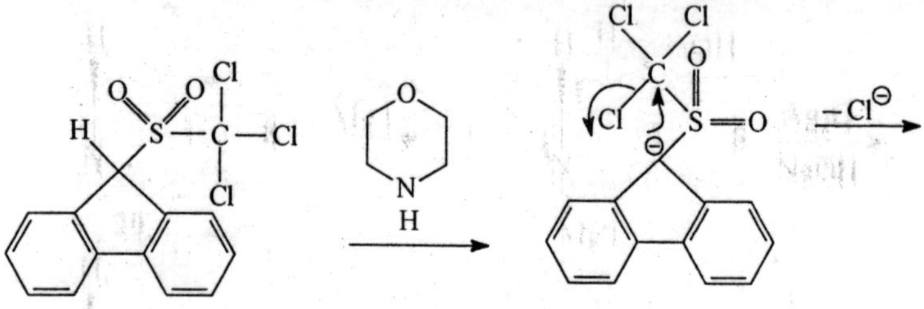

56

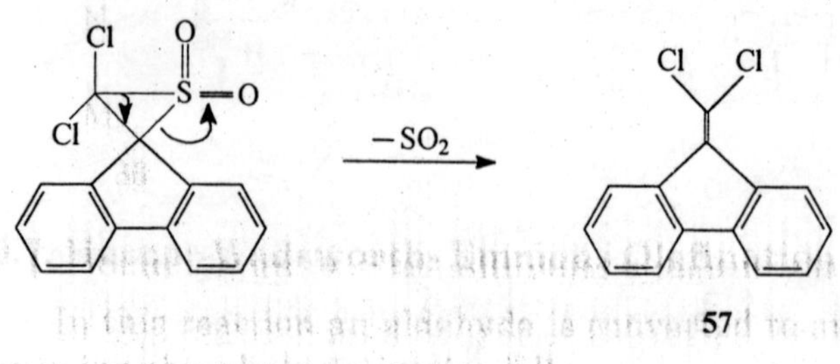

57

9.12. Shapiro Reaction

Reduction of hydrazone **58** to alkene **59** by using lithium diisopropylamide (LDA) in THF is called as shapiro reaction.[39,40]

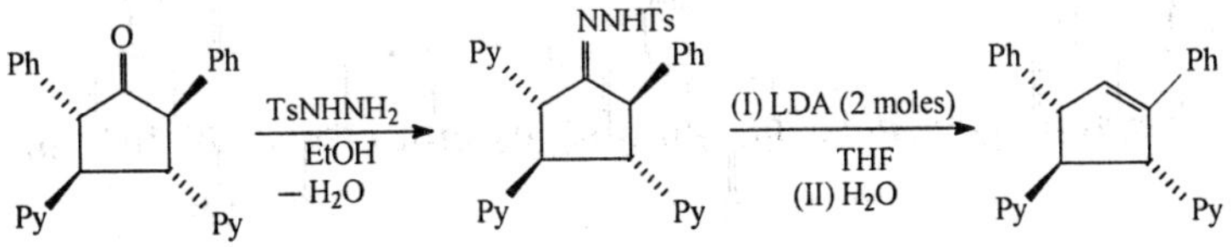

58 **59**

9.13. Sonogashira Coupling

The reaction is used for the synthesis of alkynyl alkene **60** from halo alkene and alkynes possessing terminal hydrogen. The reaction takes place in the presence of a base. The reaction involves formation of acetylide ion **61** which then couples with haloalkene to form the product **60**.[41,42]

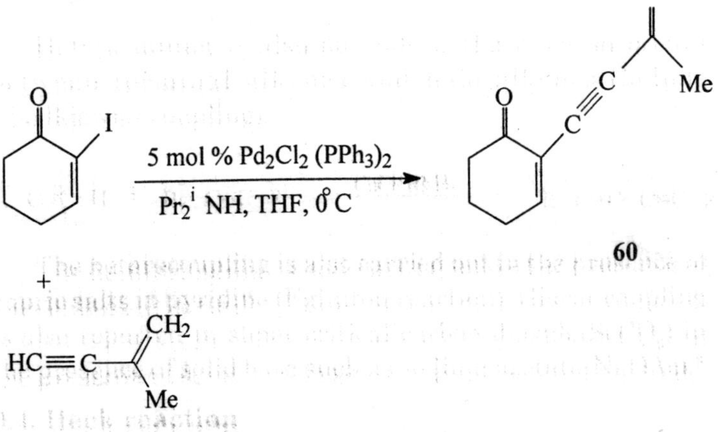

60

The mechanism of the reaction is shown below

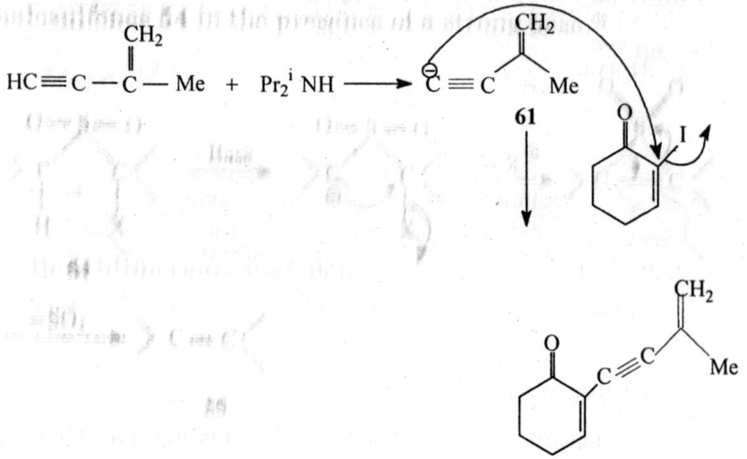

60

9.14. Wohl-Ziegler reaction

Allylic bromination of olefins with N-bromosuccinimide in the presence of peroxides or ultraviolet light is called as Wohl-Ziegler reaction.[43]

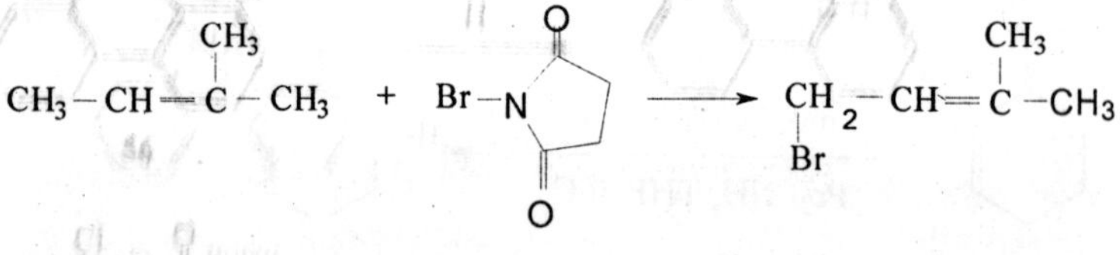

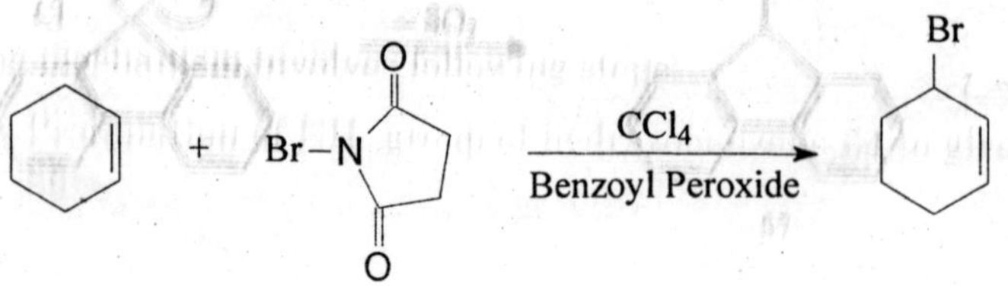

The mechanism of the reaction involves following steps:

1. Formation of free radicals through cleavage of nitrogen-bromine bond of N-bromosuccinimide.

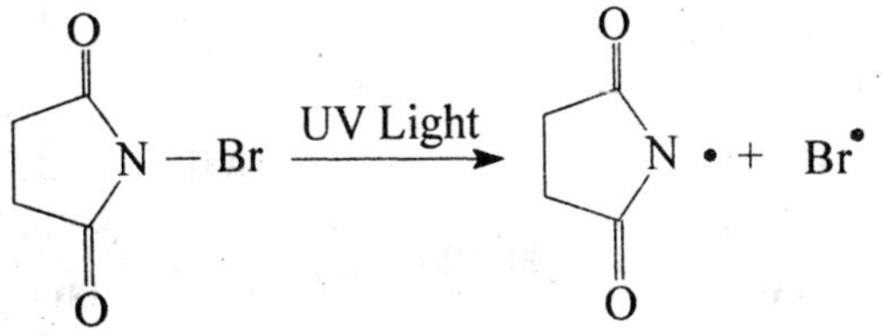

2. Removal of hydrogen free radical from allylic position to form allyl free radical which is stabilized by resonance.

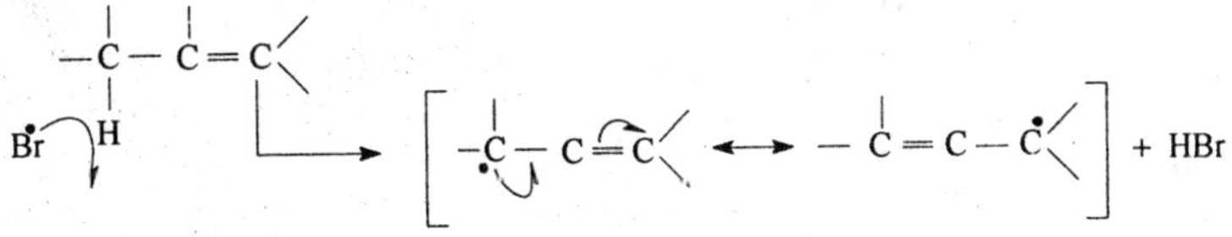

3. Coupling of N-bromosuccinimide with HBr to release bromine molecule and succinimide.

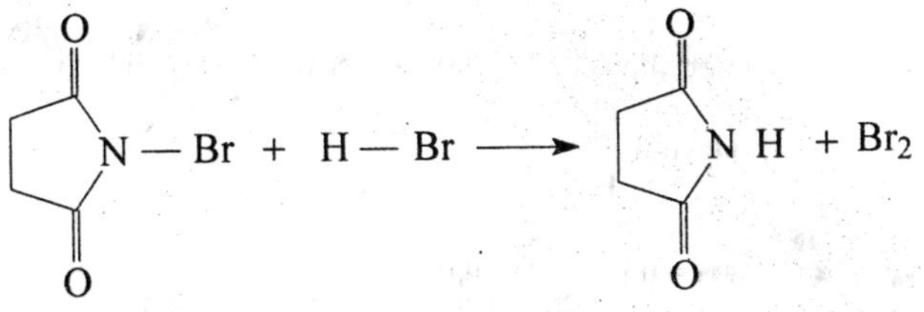

4. Coupling of alkyl free radical with bromine molecule to form alkyl bromide and bromine free radical is regenerated and chain reaction continues.

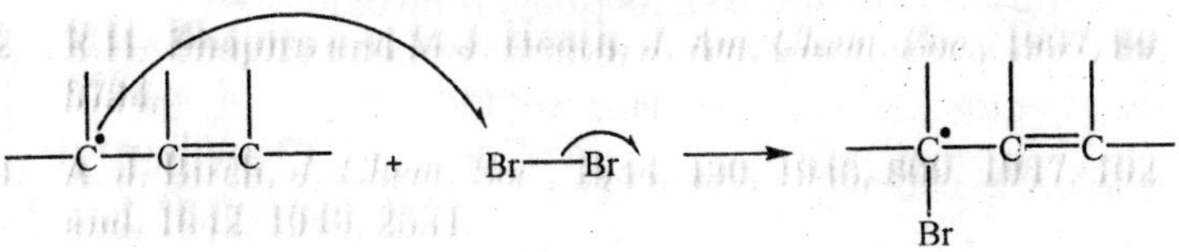

Bromination with NBS is therefore, a free radical reaction.

The advantage of NBS is (1) to provide a constant and low concentration of bromine for allylic bromination. (2) To use up the HBr liberated during the reaction. (3) Selective bromination occurs with the use of NBS.

9.15. Wittig Reaction

The reaction is used for the synthesis of alkenes from carbonyl compound (aldehydes and ketones).

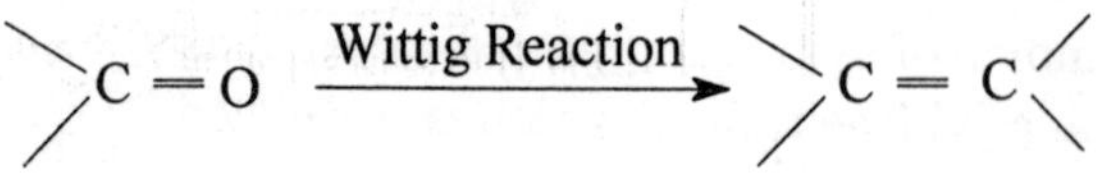

The reaction takes place in the presence of wittig reagent which is prepared from triphenylphosphine, alkylhalide (primary or secondary) and a strong base such as sodamide, alkyl lithium or aryllithium. The reaction sequences are summarized below.

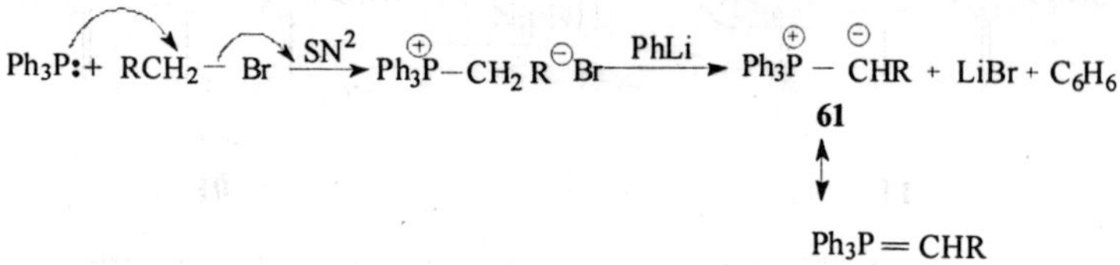

The carbanion **61** is stabilized by resonance through d orbitals of adjacent phosphonium cation. The carbanion adds to carbonyl compound to form four membered cyclic structure **62** as phosphorous has a special affinity for the carbonyl oxygen. Finally cleavage of bonds takes place and phosphorous is removed as phosphine oxide.

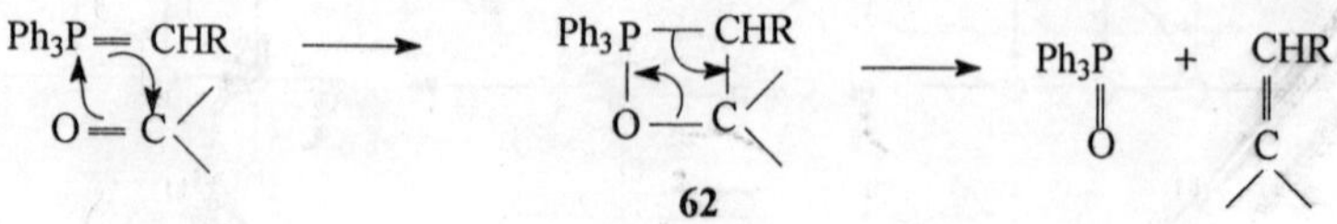

62

The reaction has been used successfully for the synthesis of 3-styrylchromones **64-65** from 3-formylchromone **63** employing benzylidene triphenyl phosphoranerane **66** as wittig reagent[44]. The reaction is summarized below.

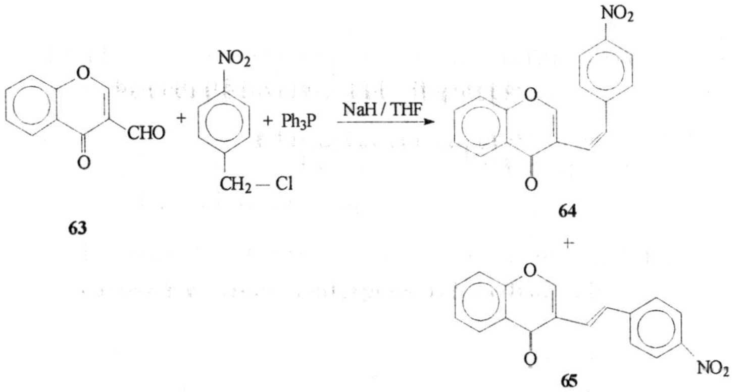

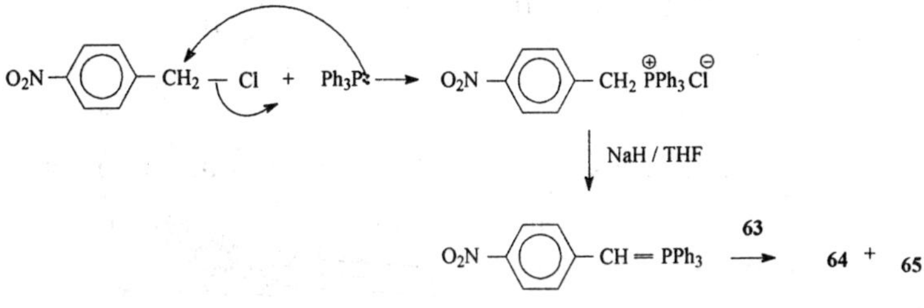

66

REFERENCES

1. W.R. Bamford and T.S Stevens, *J. Chem. Soc.*, 1952, 4735.

2. R.H. Shapiro and M.J. Heath, *J. Am. Chem. Soc.*, 1967, **89**, 5734.

3. A. J. Birch, *J. Chem. Soc.*, 1944, 430; 1946, 809, 1947, 102, and, 1642; 1949, 2531.

4. P.K. Oommen, *Aust. J. Chem.*, 1975, ***28, 2095***
5. G. S. R. Subba Rao, *Pure and App. Chem.* , 2003, **75**, 1443.
6. C. Glaser, *Ber.* , 1869, **2**, 422
7. C. Glaser, *Ann.*, 1870,**154**, 137.
8. J. Li and H. Jiang, *Chem. Commun.*, 1999,2369.
9. P. A. Patel, C. B. Zieglar, N. A. Cortese, M. A. Plevyak, C. T. Zebovity, M. Tempko, R. F. Heck, *J. Org. Chem.* , 1977, **42**, 3903.
10. N. A. Cortese, C. B. Ziegler, B. J. Hrnjez and R. F. Heck, *J. Org. Chem.*, 1978, **43**, 2952.
11. R. F. Heck, *Organic React.* 1982, **27**, 345.
12. B. C. C. Canlello, S. C. Connor, D. K. Dean and R. M. Hindley, *Synlett*, 1997, 263.
13. E. Negishi, S. Ma, T. Eughihara and Y. Noda, *J. Org. Chem.* , 1997, **62**, 1922.
14. A.W. Hofmann, *Ber.*, 1881, **14**, 659.
15. A. W. Hofmann, *Ann.* , 1851, 78, 253.
16. D. Benkes, P. Netchitailo, J. Morel and B. Decroix, *Synth. Commun.* , 1998, **28**, 949.
17. S.E. Denmark and D.S. Middleton, *J. Org. Chem.* , 1998, **63**,1604.
18. T. Murata, M. Ishikawa, R. Nishimaki and K. Tadono, *Synlett*, 1997, 1291.
19. G. E. Keck and E. P. Boden, *Tetrahedron Lett.* , 1984, **25**, 265.
20. S. Knapp and Y. Dong, *Tetrahedron Lett.*, 1997, **38**, 3813.
21. E. Knoevenagel, *Ber.*, 2596, **31**, 1898.
22. J. March, *Advanced Organic Chemistry, reaction mechanism and structure,* IIIrd Ed, John Wiley & Son.
23. A. M. Asirea, K. Ahmad, A. Abraham, F. Jalboalb, S. Zhang, *Molbank,* 2004, M359.

24. A. Dandia, P. Sarawgi, K. Arya and S. Khaturia, *Arkivoc* , 2006, **(xvi),** 83.

25. D. L. Levesque, E. C. Wang, D. C. Wei, C. C. Tzeng, R. P. Panzica, F. N. M. Naguib and M. H. el Kouni, *J. Heterocyclic Chem.* , 1993, **30**, 1399.

26. K. N. Ali and H. Aboulghasem, *Iran J. Chem .and Chem. Eng.*, 2001, **20** (1), 9.

27. X. Zhang, E. S. M. Lai, R. Martin- Aranda and K. L. Yeung *App. Catalysis A: General*, 2004, **261**, 109.

28. J. S. Yadav, B. S. S. Reddy, A. K. Basak, B. Visali, A. V. Narsaiah, K. Nagaiah, Eur. *J. Org. Chem.*, 2004, 546.

29. S. R. Mahalle, P. D. Netankar, S. P. Bondge and R. A. Mane, *Green Chem. Lett. and Rev.*, 2008, **1(2),** 103.

30. G. W. Wang and B. Cheng, *Arkivoc*, 2003, (x), 4.

31. D. Prajapati and M. Gohain, *Beils. J. Org. Chem.* , 2006, **2**, 11.

32. C. H-Guzman, P, I-Vasquez, G. Z-Torres and B. K. Cassels. *J. Chil. Chem. Soc.*, 2004, **49 (3**), 257.

33. D. Prajapati and K. H. J. Borah, *Beil. J. Org. Chem.*, 2007, **3 (43)**, doi:10.1186/1860-5397-3-43.

34. M. T. Thorat, M. H. Jagdale, R. B. Mane, M. M. Salunkhe and P. P. Wadagaonkar,.*Curr. Sci.*, 1987, **56**, (15), 771.

35. T. Eguchi, K. Ibaragi and K. Kakinuma, *J. Org. Chem.* ,1998, **63**, 2689

36. E. Lee, and C. H. Yoon, *Tetrahedron Lett.* ,1996, **37**, 5929.

37. L. Ramberg and B. Backlund, *Arkiv Kemi Mineral Geol.* , 1940, **13A**, 1940.

38. S. Braverman and Y. Zafrani, *Tetrahedron*, 1998**, 54,** 1901.

39. R.H. Shapiro, *Org. React.*, 1976**, 23**, 405.

40. U. Siemeling, B. Neumann and H-G. Stammler, *J. Org. Chem.*, 1997, **62**, 3407.

41. K. Sonogashira, Y. Tohda and N Hagihara, *Tetrahedron Lett.*, 1975, 4467.

42. M.W. Miller and C.R. Johnson, *J. Org. Chem.*, 1997, **62**, 1582.

43. A. Wohl, *Ber.*, 1919, 52, 59, L. Ziegler et al., *Ann.*, 1942, **551**, 30.

44. A. Sandulache, A.M.S. Silva, D.C.G.A. Pinto, L.M.P.M. Almeida and J.A.S. Cavaleiro, *New J. Chem.*, 2003, **27**, 1592.

10

Synthesis of Alcohols / Phenols

10.1. Barton reaction

The reaction consists in treating an alcohol **1** with nitrosylchloride to give nitrile ester **2** which upon photolysis is converted to γ-oximino alcohol **3**.[1] The reaction has been successfully carried out with 18-substituted steroids where a methyl group in the γ position to an OH group is converted to oxime group.

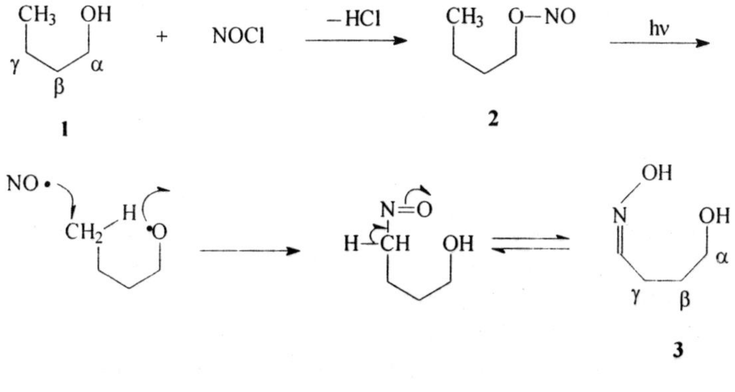

10.2. Bouveault Blanc reaction

Reduction of esters to primary alcohols by means of

sodium in the presence of a proton donor solvent is called as Bouveault Blanc reduction.[2]

$$R{-}\overset{O}{\overset{\|}{C}}{-}OR' \xrightarrow[\text{Alcohol}]{\text{Na +}} RCH_2OH + R'OH$$

Ester **4** is first reduced to aldehyde **8** which is further reduced to primary alcohol **9**.

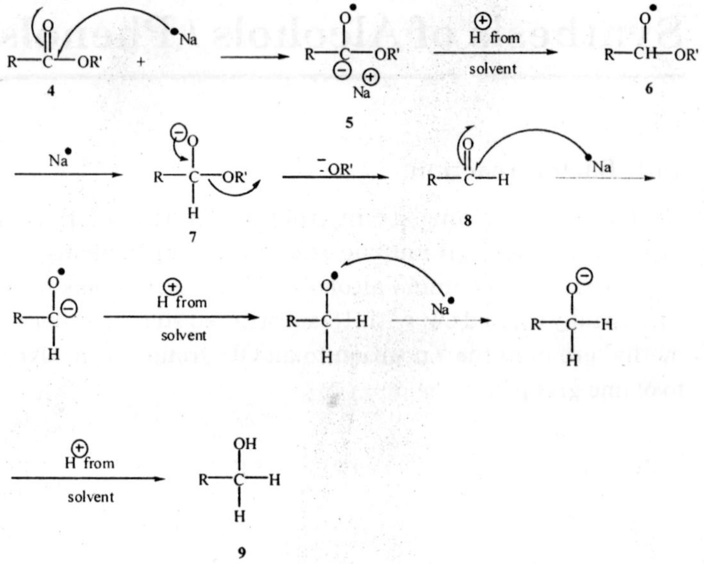

Formation of aldehyde **8** from ester involves electron donation from sodium metal to carbonyl group of ester to form radical anion **5**. The radical anion **5** takes a proton from solvent (alcohol) to give radical **6**. The radical **6** react with another molecule of sodium to form anion **7** which is then converted to aldehyde **8** . The reduction of aldehyde **8** with sodium and alcohol in similar manner forms alcohol **9**.

10.3. Cannizzaro reaction

Self oxidation-reduction reaction of aldehyde containing

no α-hydrogen for enolisation in the presence of concentrated alkali solution to give alcohol and salt of corresponding carboxylic acid is called as cannizzaro reaction.[3] The carboxylic acid is regenerated from its salt upon acidification. The reaction between two different aldehydes containing no α-hydrogen in the presence of alkali is called as crossed cannizzaro reaction. If an alkaloxide is used in the place of alkali solution then an ester is obtained along with an alcohol.

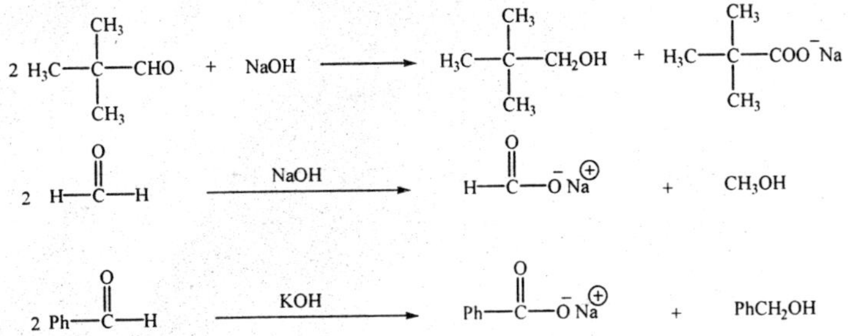

In the reaction one molecule of aldehyde acts as a hydride donor and another acts as accepter. The reactions is initiated by attack of hydroxide ion on carbonyl carbon to form anion **10** which then donates hydride ion to another molecule of aldehyde to form alcohol **11** and salt of carboxylic acid **12**.

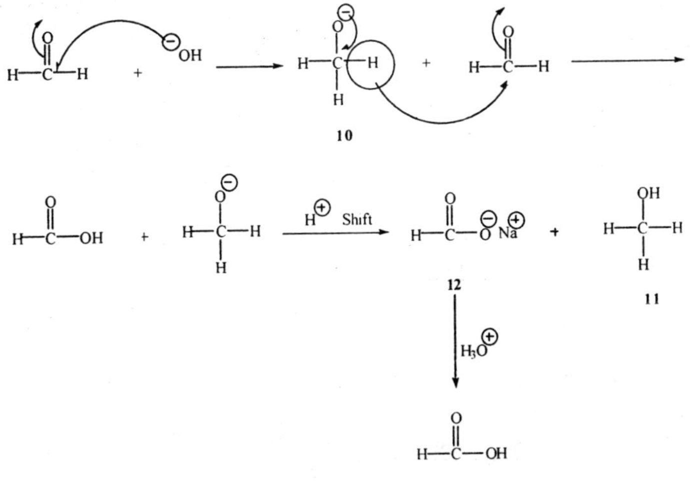

In crossed cannizzaro reaction a mixture of all possible products are obtained. However, if one of the aldehyde is formaldehyde then due to higher reactivity towards nucleophilic addition reaction, formaldehyde preferably undergoes nucleophilic addition by hydroxide ion and acts as hydride donor. The result is, therefore, formation of sodium formate along with alcohol.

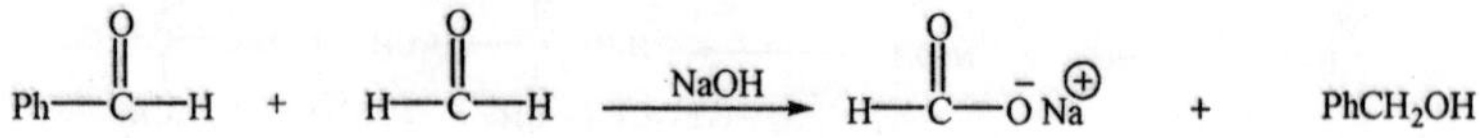

If two aldehydic groups are present in the same molecule (having no α hydrogen) then intramolecular cannizzaro reaction takes place in the presence of base. Thus, glyoxal **13** forms hydroxyacetic acid **14**.

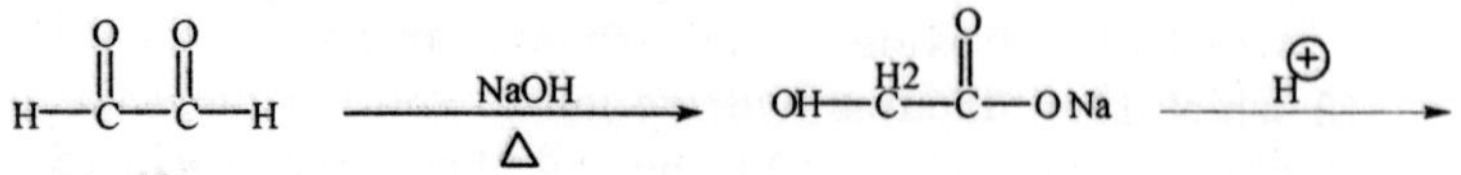

13

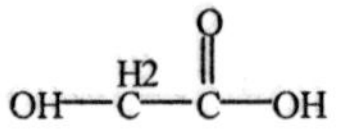

14

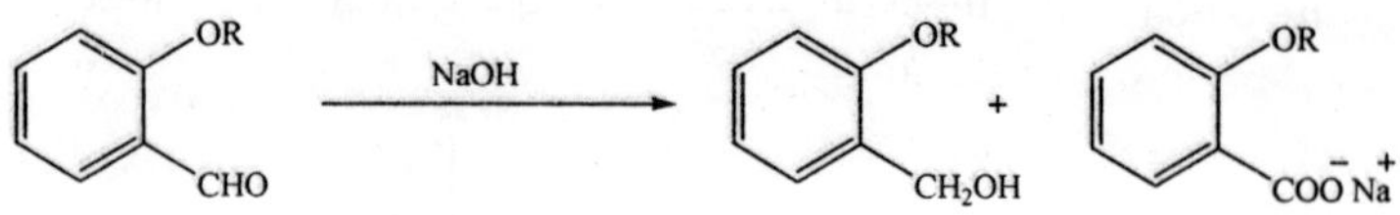

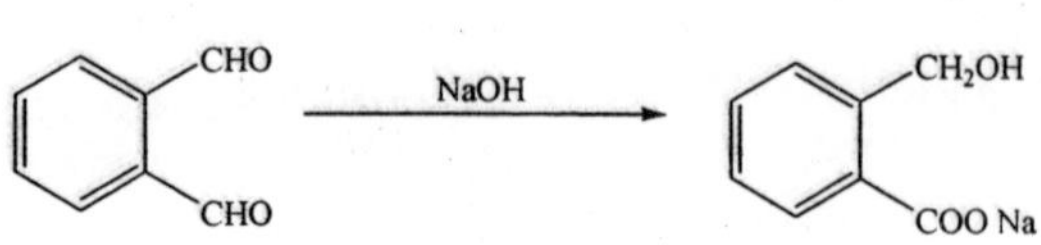

10.4. Favoraski reactions

The reaction is named after its Russian discoverer A. Y. Favoraski. The reaction involves nucleophilic attack of a terminal alkyne **15** on carbonyl carbon of a ketone **16** in the presence of a base.

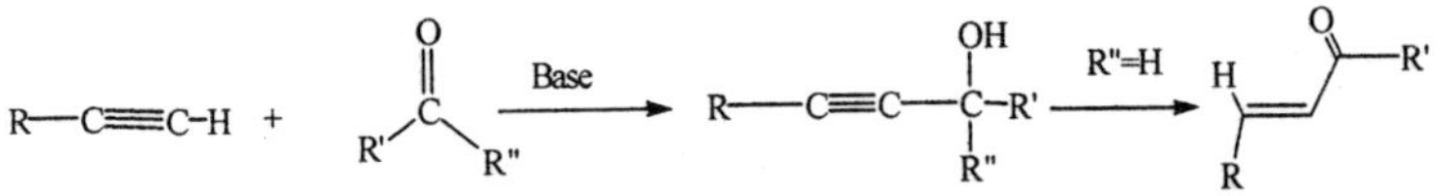

15 16 17

If ketone contain an α hydrogen then tautomerization occurs and enone **17** is obtained.

The mechanism involves formation of acetylide in the presence of a base.

$$R—C\equiv C—H + NaNH_2 \xrightarrow{-NH_3} R—C\equiv C^{\ominus}\ Na^{\oplus}$$

Acetylide ion then reacts with aldehyde/ ketone to form alcohol **18**.

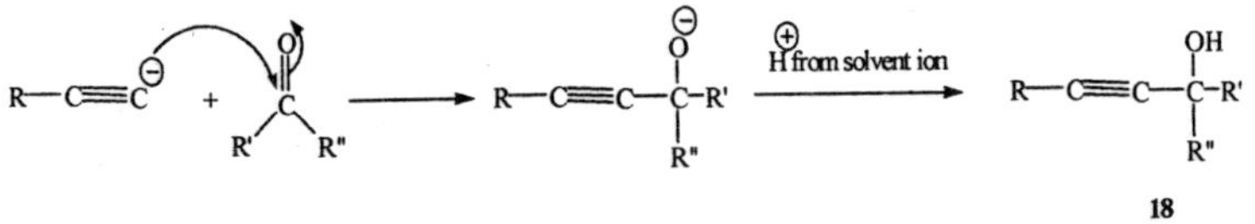

18

10.5. Wolffenstein-Boters reaction

The reaction involves oxidation nitration of aromatic compounds to give nitrophenols upon treatment with nitric acid or higher oxides of nitrogen. The reaction is catalyzed by mercury salts such as mercuric nitrate. Thus, picric acid is obtained in quantitative yield from benzene.[5]

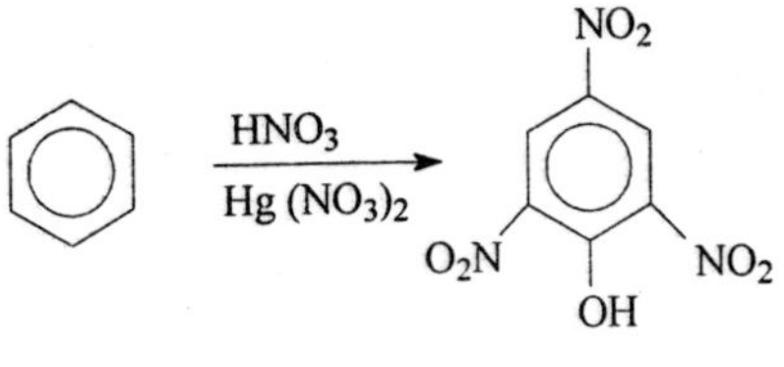

REFERENCES

1. B.H.R. Barton et al. *J. Am. Chem. Soc.*, 1960, **8**, 2640; 1961, **83**, 4076.
2. L. Bouveault, G. Blanc. *Compt. Rend.*, 1903, **136**, 1676; *Bull. Soc. Chim. France*, 1904, **31 (3)**, 666.
3. S. Cannizzaro, *Ann.*, 1853, 88, 129; K. List and H. Limpricht, *Ann.*, 1854, 90, 180.
4. R. Wolffenstein and O. Boters, *Ber.*, 1913, **46**, 586.

11

Synthesis of Miscellaneous Compounds

11.1. Acetoacetic ester synthesis

Ethylacetoacetate **1** is the ethylester of acetoacetic acid **2**. It is considered as an active methylene compound and is obtained by Claisen condensation of ethylacetate **3** in the presence of a strong base such as sodium ethoxide.

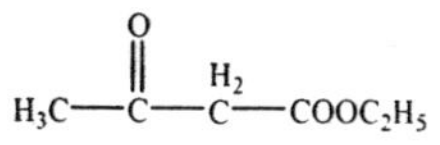

1

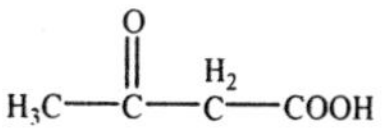

2

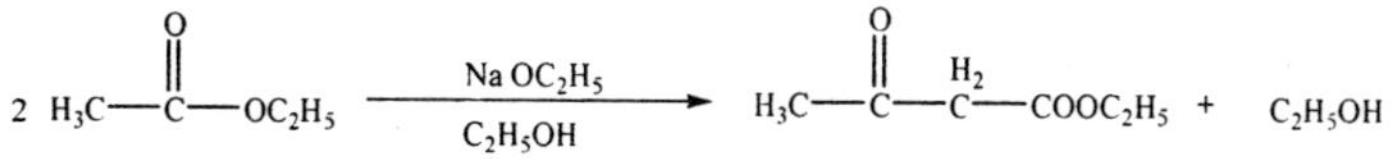

3 1

The reaction involves following steps.

1. Formation of carbanion **4** by removal of an α H from **3** which is stabilized by resonance.

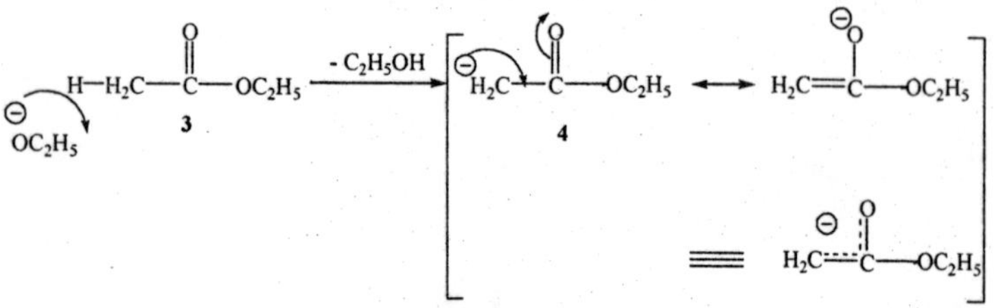

3. Attack of carbanion **4** on carbonyl carbon of second molecule of **3** followed by expulsion of $^{-}OC_2H_5$ from the resulting molecule to give **5**.

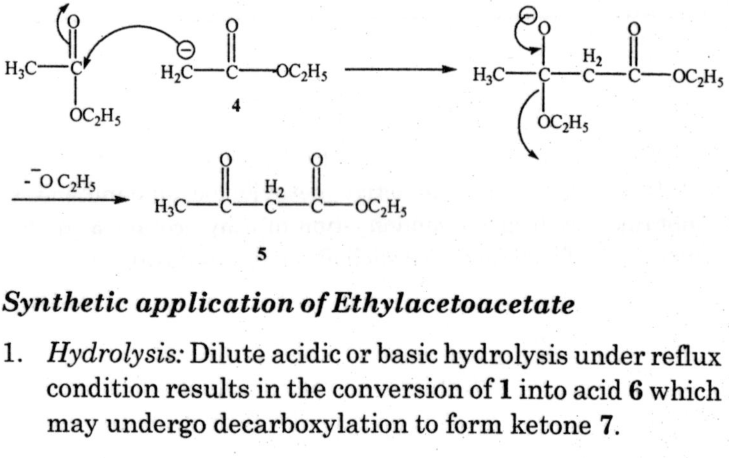

Synthetic application of Ethylacetoacetate

1. *Hydrolysis:* Dilute acidic or basic hydrolysis under reflux condition results in the conversion of **1** into acid **6** which may undergo decarboxylation to form ketone **7**.

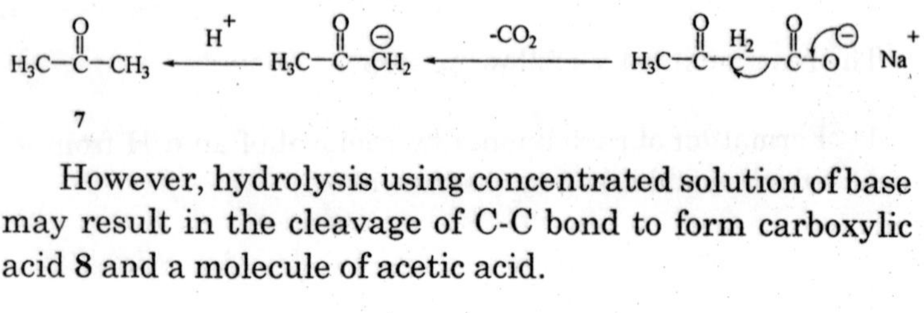

However, hydrolysis using concentrated solution of base may result in the cleavage of C-C bond to form carboxylic acid **8** and a molecule of acetic acid.

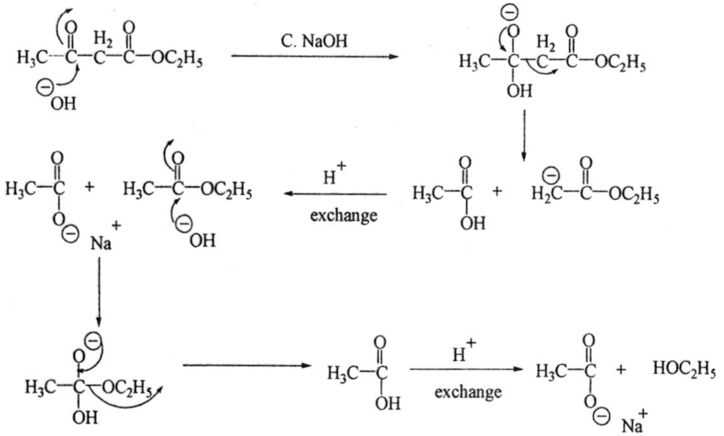

8

Thus, we can synthesize alkyl/aryl/acyl substituted carboxylic acids as well as ketones by performing hydrolysis under different reaction condition e.g.

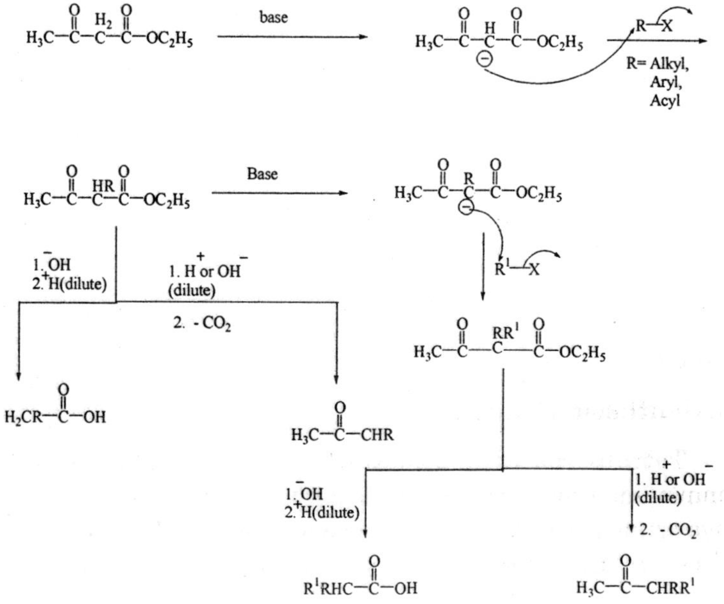

(2) Knoevenagel condensation reaction between benzaldehyde and ethylacetoacetate in microreactor using Cs-exchanged NaX catalyst and NaA membrane was reported by Yeung et.al.[1]

By using thin CsNaX film in place of powdered catalyst the selectivity and yield of the product was increased and the formation of side products was minimized.

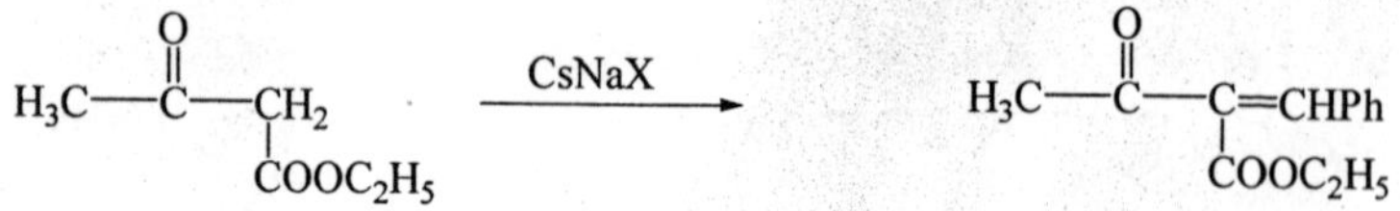

3. Synthesis of pyrazolones

Synthesis of various pyrazolones **9** from ethyl aceto acetate and hydrazine have been reported.[2]

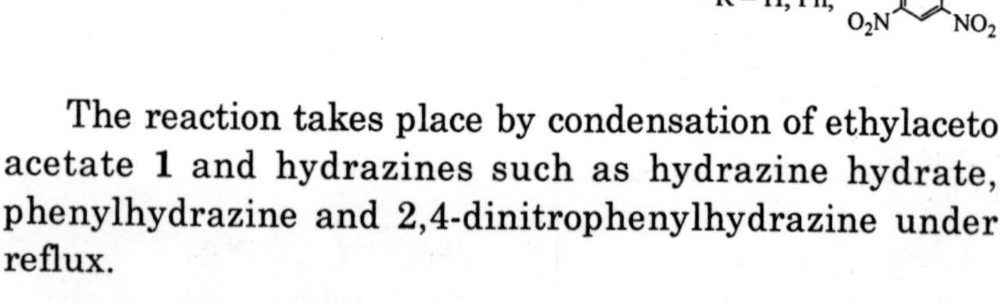

The reaction takes place by condensation of ethylaceto acetate **1** and hydrazines such as hydrazine hydrate, phenylhydrazine and 2,4-dinitrophenylhydrazine under reflux.

4. Synthesis of Indazoles

Tetrahydro-1H-indazoles **12** have been synthesized employing ethyl acetoacetate **1**, 2-hydroxy chalcone **10** and hydrazine hydrate under microwave irradiation conditions.[3] First an intermediate **11** is obtained by the Michael condensation of ethyl acetoacetate and hydroxychalcone **10** in the presence of K_2CO_3.

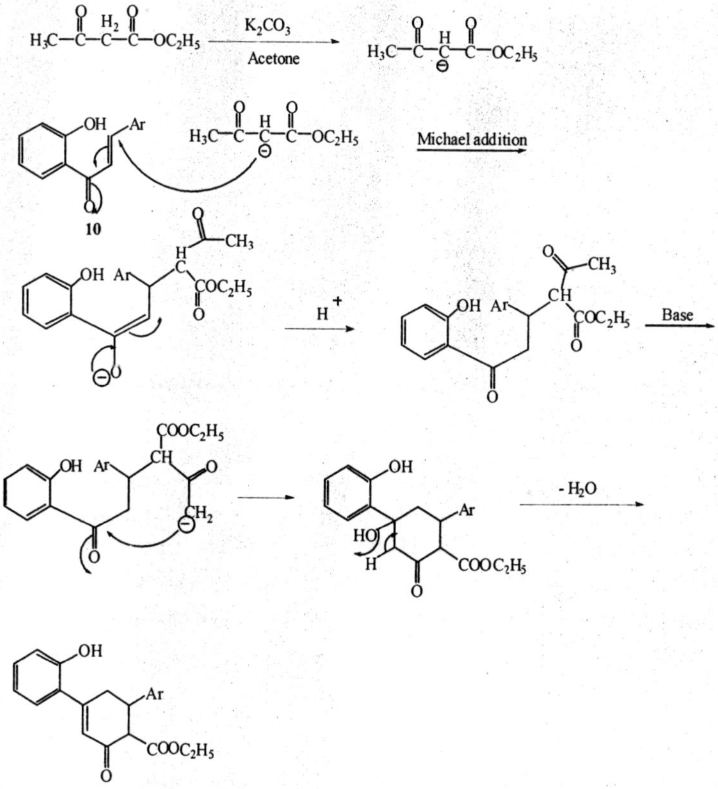

11

Treatment of cyclohexanone intermediate **11** with hydrazine hydrate forms indazole **12**.

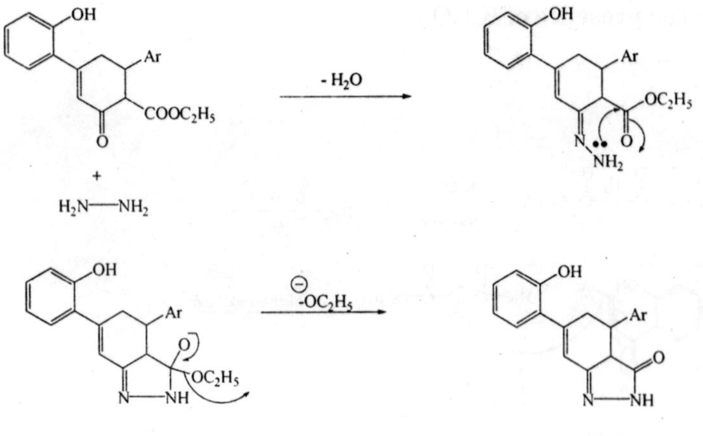

12

5. **Synthesis of dihydropyrimidones and imidazolene-2-ones.**

Novel dihydropyrimidinones **13** are synthesized employing alkyl aceto acetate, phenylglyoxal and urea in the presence of $ZnCl_2$. However, imidazoline-2-one **14** derivatives are obtained when condensation of alkyl acetoacetate, phenylglyoxal is carried out with methyl urea.[4]

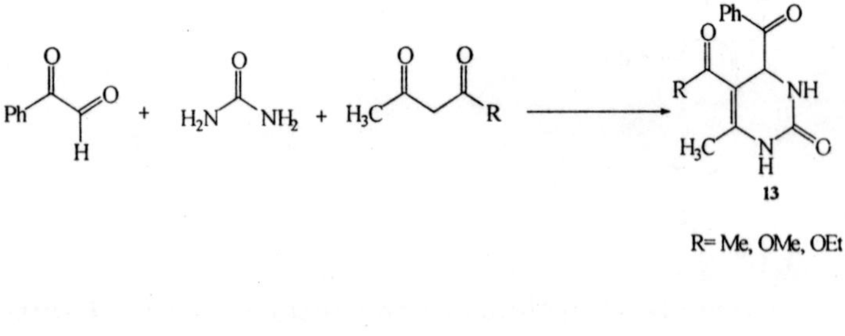

13

R= Me, OMe, OEt

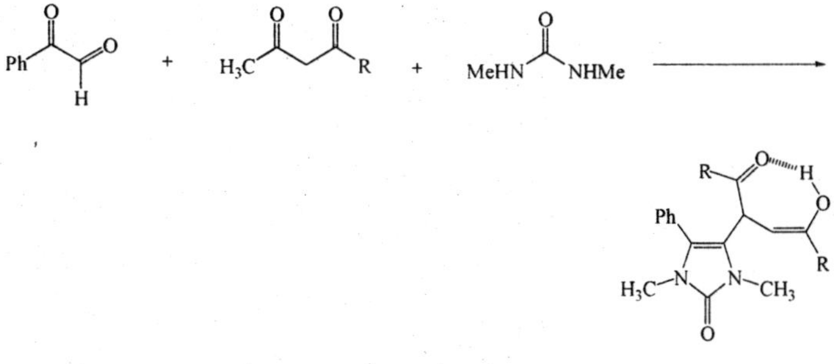

14

A combinatorial chemistry approaches has been used for the formation of these products.

6. Synthesis of polyhydroquinolines

Polyhydroquinoline derivatives **15** have been synthesized employing ethylaceto acetate benzaldehyde, dimedone and ammonium acetate at room temperature in the presence of catalyst MCM-41 under solvent free condition.[5]

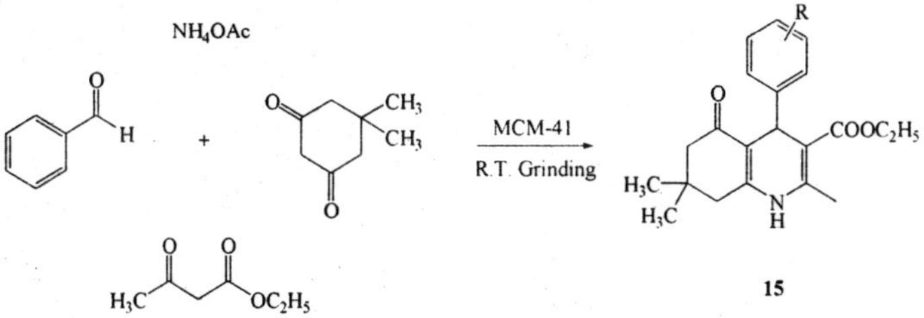

15

The product **15** is obtained via Hantzsch condensation in the presence of heterogenous catalyst, MCM-41.

7. Synthesis of Pyrazolones

A variety of pyrazolones **16** have been synthesized under microwave irradiation conditions employing ethylaceto acetate/ ethyl benzoyl acetate and various kinds of hydrazines.[6]

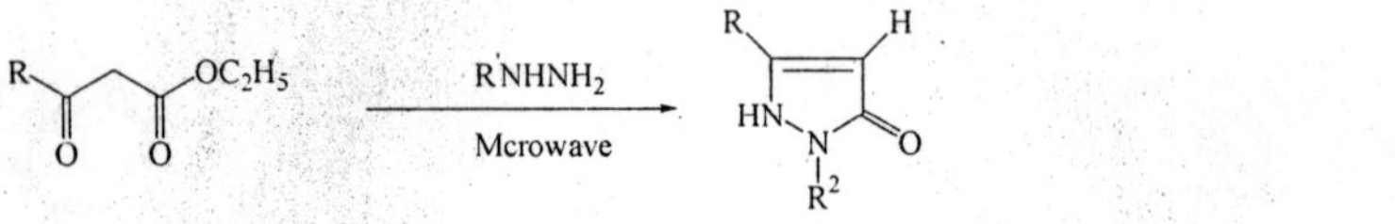

16

R= CH_3, Ph

R^2=H, Ph, $CONH_2$ 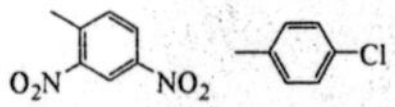

8. Synthesized of enaminone derivatives

Various kinds of enaminone derivatives **17** have been synthesized using ethyl or benzyl aceto acetate (also known as β-keto ester) and amines at room temperature in the presence of Lewis catalyst, $LaCl_3$.[7]

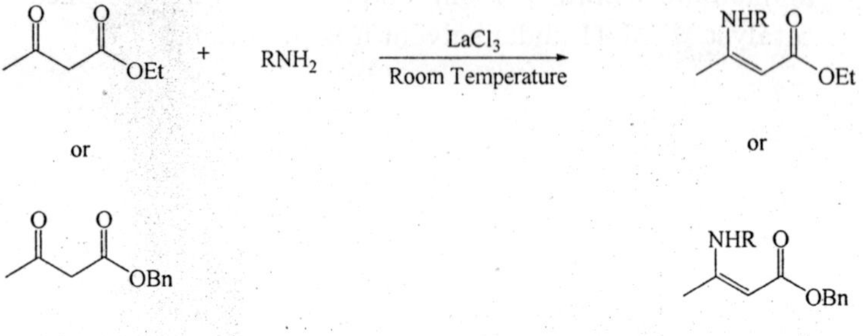

17

9. Synthesis of β-arylamino crotonates

β-Arylamino crotonates **18**, used as an intermediate for the synthesis of 2,4-disubstituted quinolines, have been synthesized under microwave irradiation employing ethylaceto acetate and aryl amine in the presence of inorganic solid supports such as SiO_2, Al_2O_3 (acidic, neutral, basic) and molecular sieve.[8]

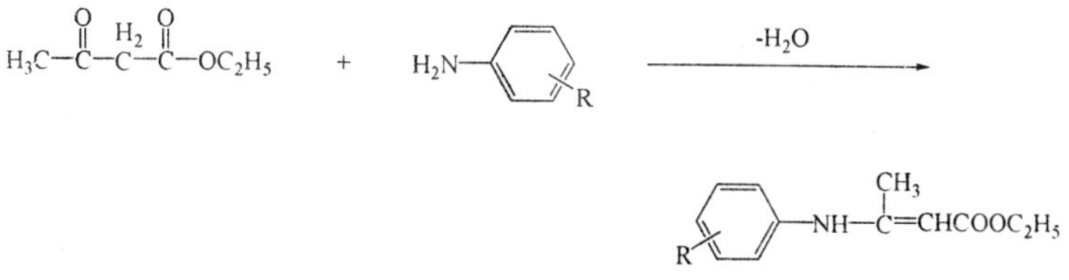

18

10. Synthesis of amino acid ester

Ester of amino acids have been prepared using ethylacetoacetate and benzyl or picolyl halide in the presence of strong base such as N,N,N', N'-tetramethyl guanidine in solvent dimethyl formamide.[9]

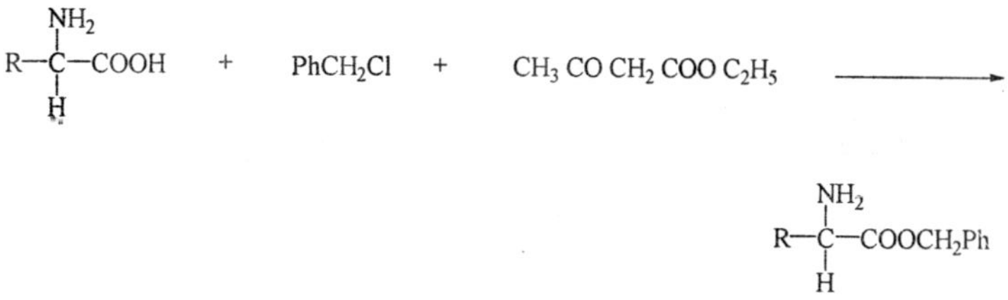

11.2. Arens-van Dorp reaction

This method is used for the synthesis of alkoxy ethynyl alcohols **21** from a ketone **19** and ethoxyacetylene **20**.[10,12]

1\.

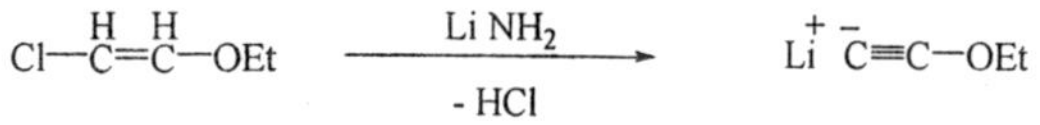

2\.

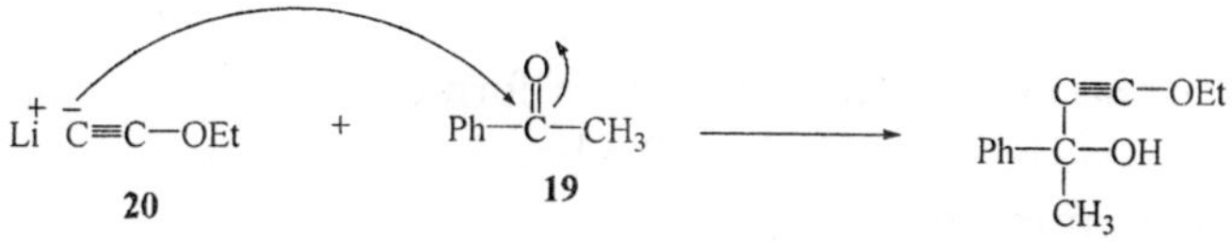

21

The reaction involves treatment of α-chloro vinyl ethyl ether with lithium amide to form lithium ethoxyacetylene which is then condensed with a ketone to form alcohol **21**.

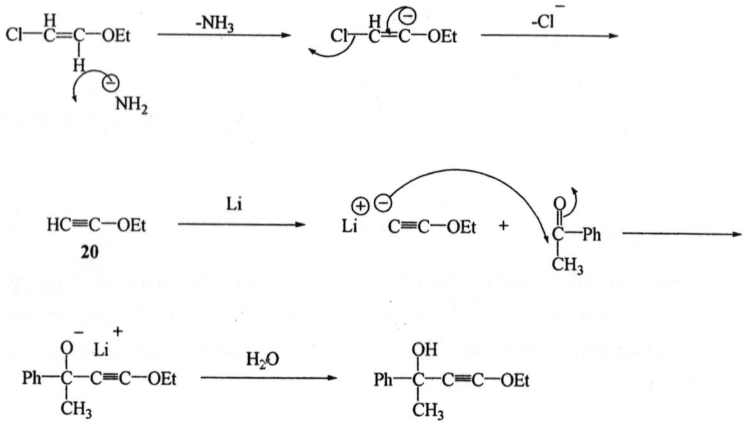

11.3. Bucherer-Bergs Reaction

The reaction is used for the synthesis of hydantoins **22** using a ketone, potassium cyanide and ammonium carbonate in a multicomponent system.

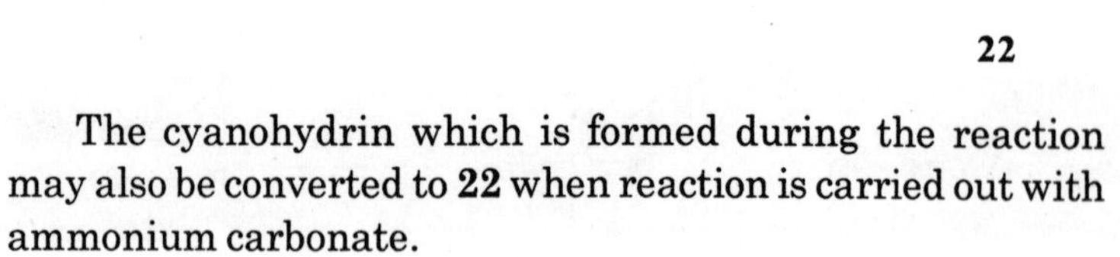

22

The cyanohydrin which is formed during the reaction may also be converted to **22** when reaction is carried out with ammonium carbonate.

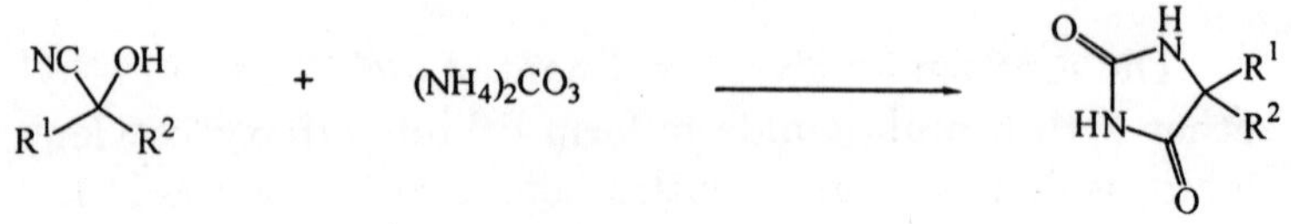

The mechanism of the reaction involves following steps.

1. Formation of cyanohydrin **23** by nucleophilic addition of cyanide ion carbonyl carbon.

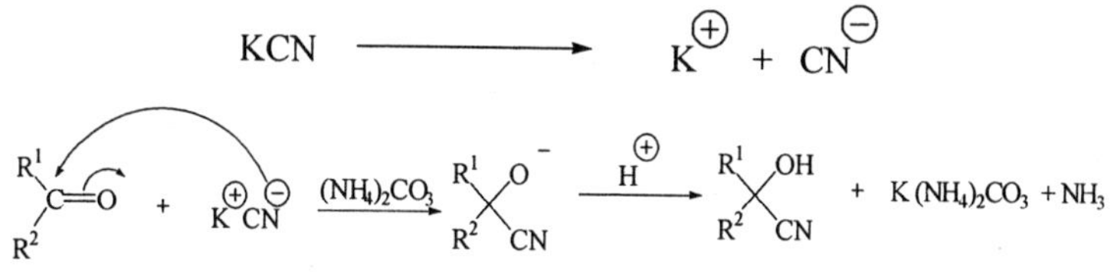

23

2. Reaction of ammonia with cyanohydrin **23** to give dehydrated product **24**.

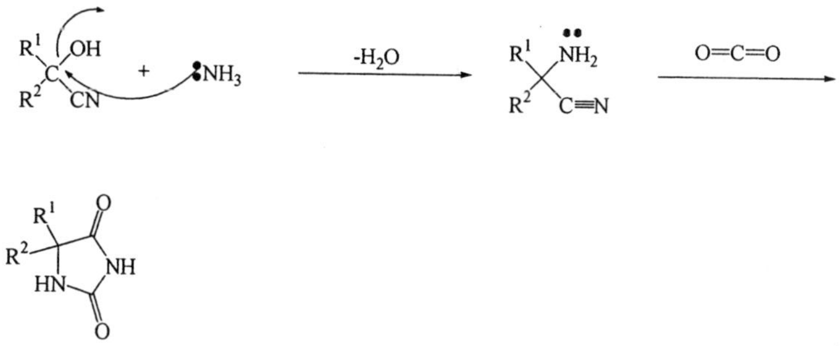

24

Pyranopyridine **25** has been converted to hydantoin derivative **26** successfully using KCN and ammonium carbonate.[13]

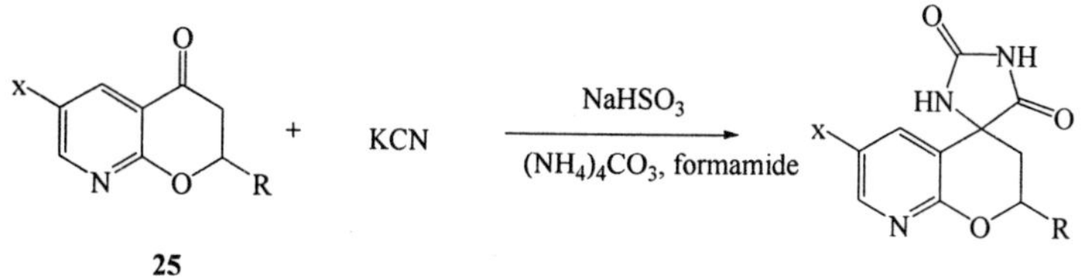

25 **26**

A modified Bucherer-Bergs reaction involves alkyl cyanide and organometallic compounds to form imine derivative which then reacts with KCN and ammonium carbonate to form hydantoin.[14]

11.4. Clemmensen reduction

The reaction is used for the synthesis of hydrocarbons from carbonyl compounds. The reaction is carried out in the presence of amalgamated zinc and hydrochloric acid.[15]

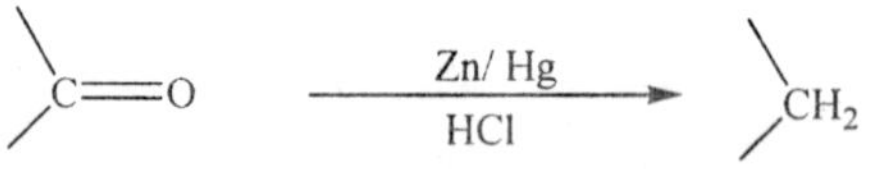

The mechanism of the reaction is described below.

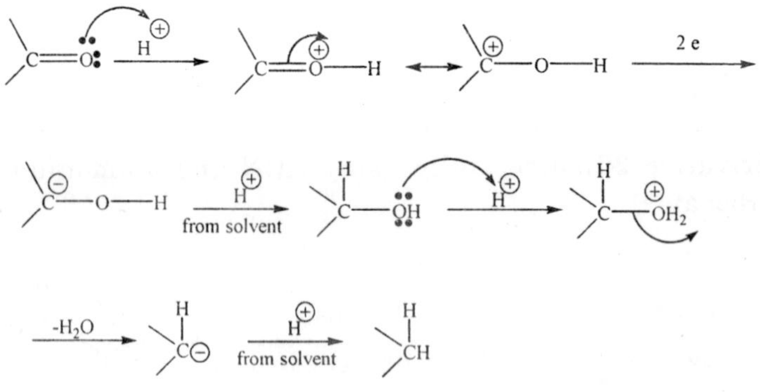

11.5. Friedel Craft Acylation

The reaction involves alkylation or acylation of aromatic compounds in the presence of anhydrous $AlCl_3$.[16-18] Other Lewis catalyst may also be used as catalyst in this reaction.

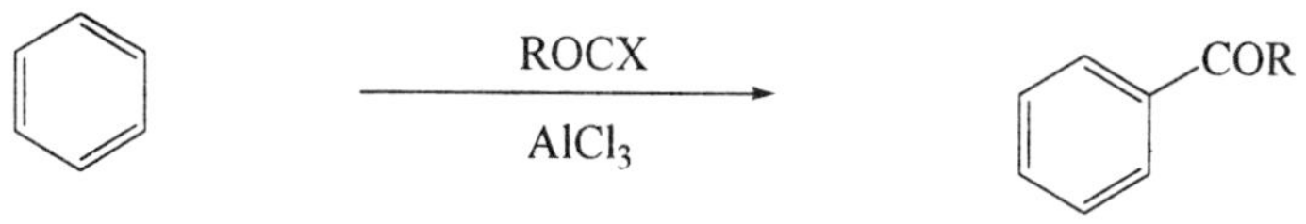

Where RCOX= Acylhalide, anhydrides

Thus, hydrocarbon **27** has been converted to **28** by refluxing with $AlCl_3$ in dichloromethane.[19]

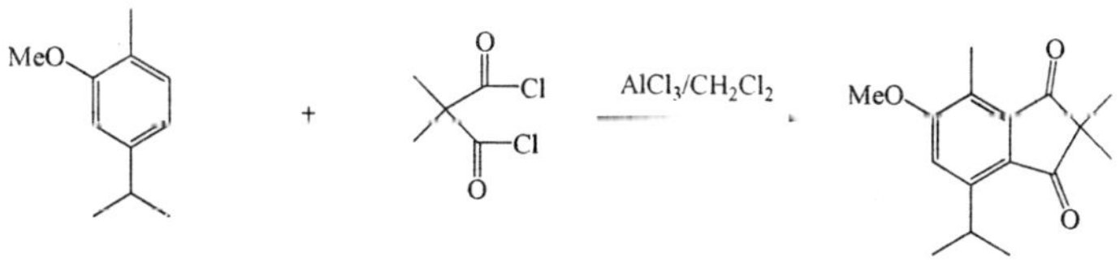

27

28

11.6.Helferich reaction

The reaction consists in heating an acetylated sugar with phenol in the presence of a Lewis acid such as $ZnCl_2$, $FeCl_3$ or p-toluene sulfonic acid.[20]

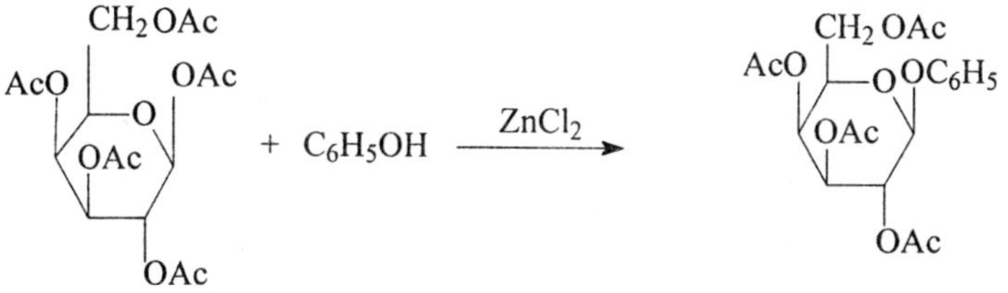

11.7.Henry reaction

Nitroalkanes having α hydrogen (carbon,possessing hydrogen, situated adjacent to nitro group) undergo condensation reaction with carbonyl compounds (aldehydes and ketones) to give condensed product, β-hydroxy nitro compound **29** by bases (similar to aldol).[21] This reaction is also known as nitroaldol condensation reaction.

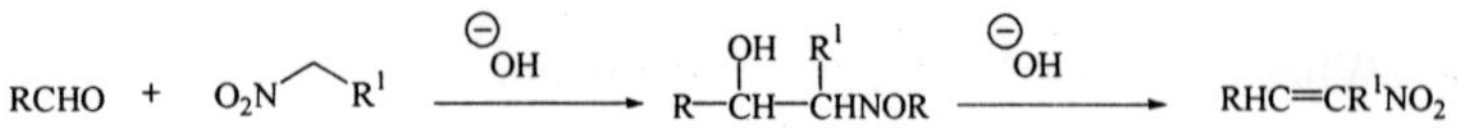

29 **30**

β-hydroxynitro derivative may dehydrate to give α, β-unsaturated nitro compound **30**.

The mechanism of the reaction is depicted below.

The following steps are involved.

1. Formation of carbanion **31** when nitro compound possessing α H is heated with a base. The formation of carbanion **31** is feasible as it is stabilized by the resonance.

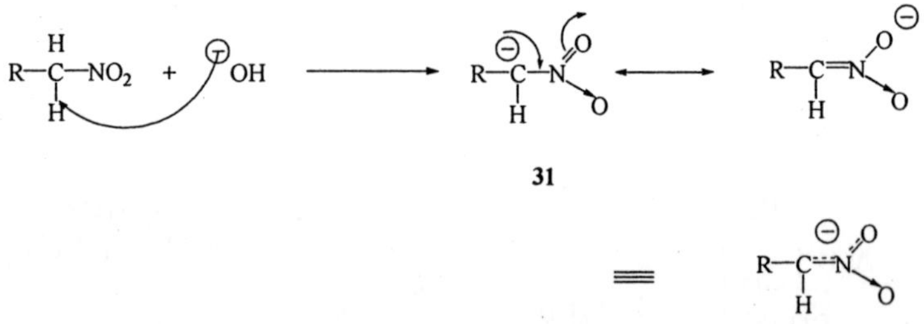

31

2. The carbanion attacks on carbonyl carbon of an aldehyde or ketone to form aldol type of product **29** which undergoes dehydration to form unsaturated compound **30**.

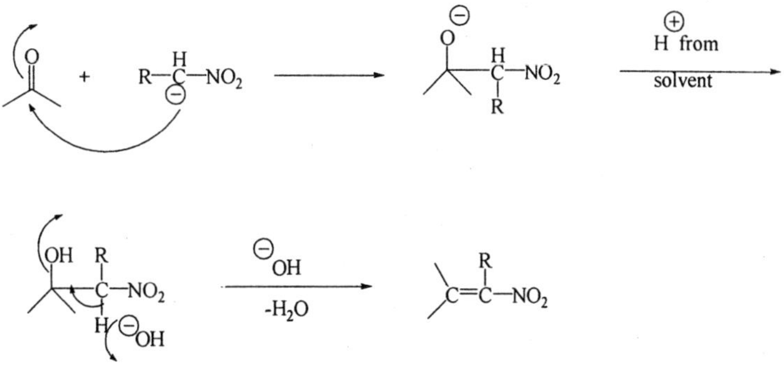

29 **30**

Synthetic importance:- Synthesis of furan derivatives **34** has been carried out employing furfural **32**, nitrobutane **33** in the presence of base.[22]

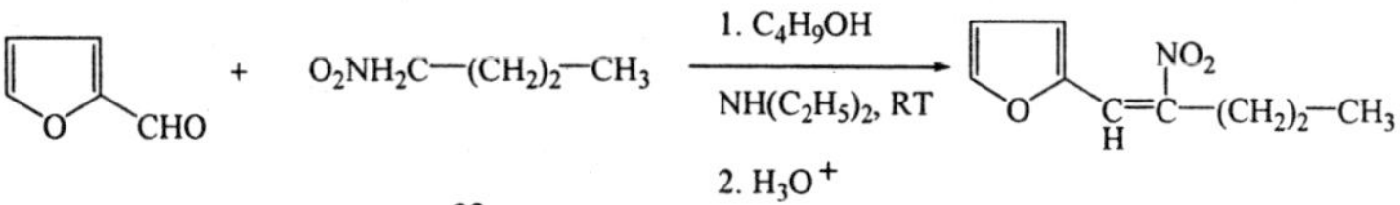

32 **33** **34**

Nitrostilbene **35** is conveniently synthesized from benzaldehyde under microwave irradiation conditions employing nitroethane and ammonium acetate.[23]

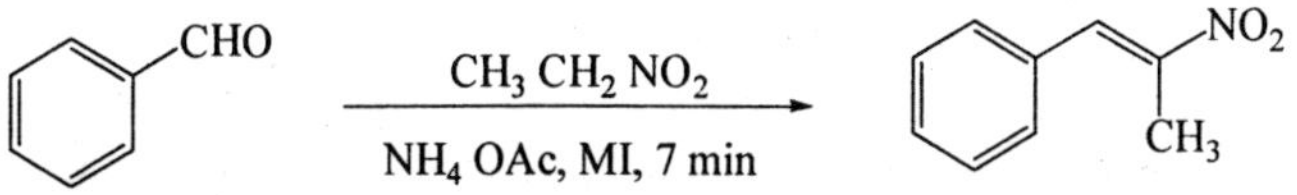

35

11.8. Hinsberg method for synthesis of thiophenes

This is a convenient method for synthesizing a wide variety of thiophene derivatives. The reaction consists in heating 1,2 dicarbonyl comounds with diethyl thiodiacetate in the presence of base.[24]

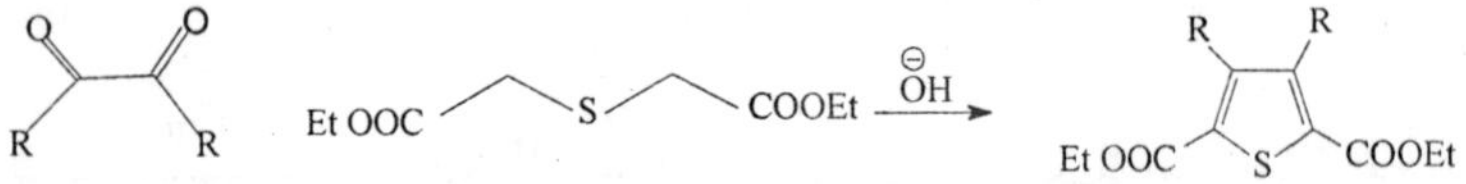

The mechanism of above reaction is described below.

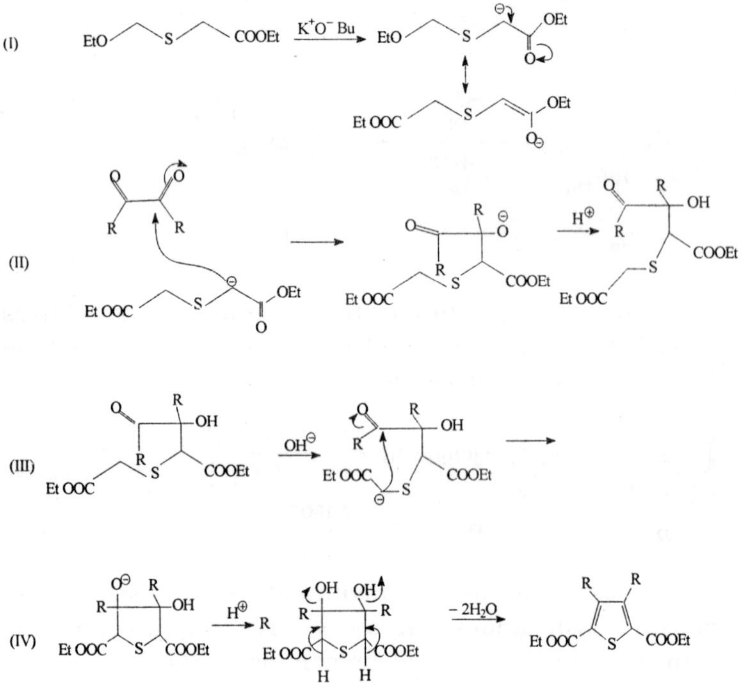

11.9. Luche reduction

Reduction of aldehydes **36** and ketones **38** to primary alcohl, **37** and secondary alcohols **39** with sodium borohydride in the presence of $CeCl_3.7H_2O$ is known as Luche reduction.[25-27]

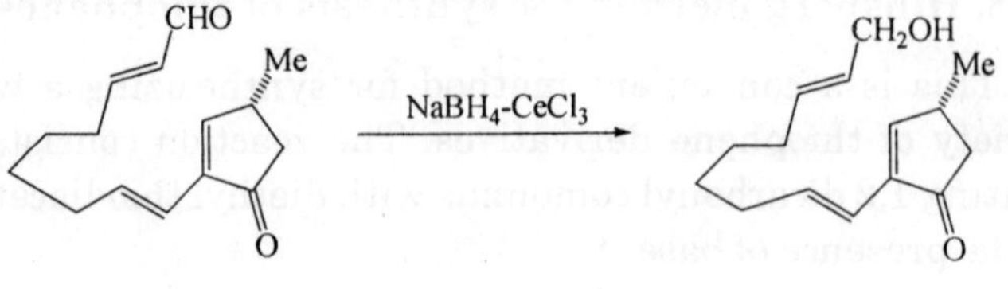

36 **37**

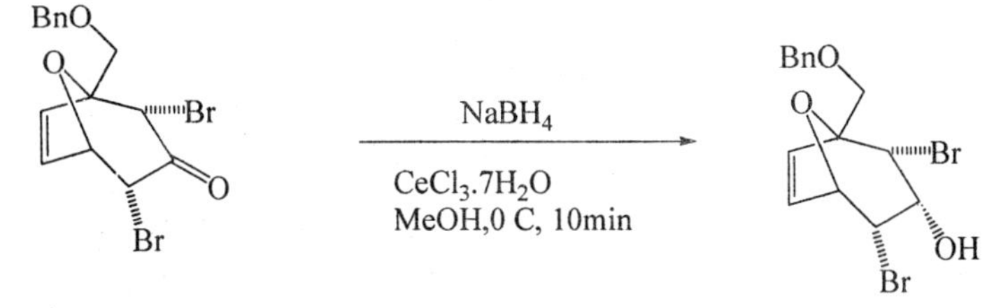

38 **39**

11.10. Mitsunobu reaction

The conversion of an alcohol **40** to tertiary amine **41** is called as Mitsunobu reaction.[28-30]

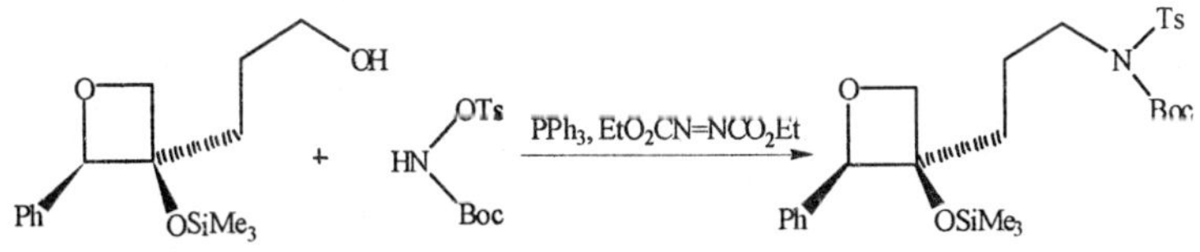

Secondary alcohol **42** has been also transformed successfully to amine **43**[31]

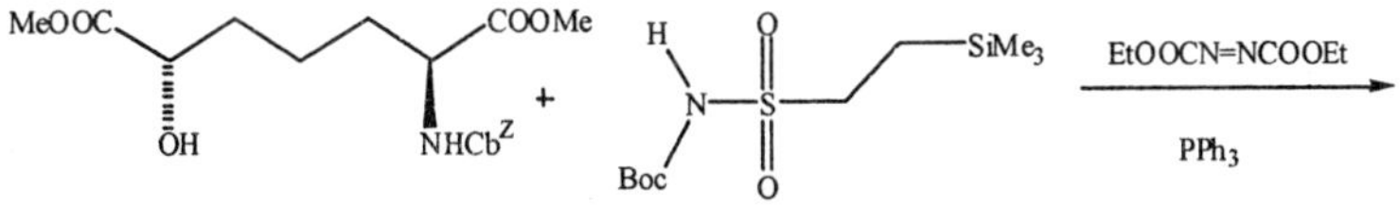

42

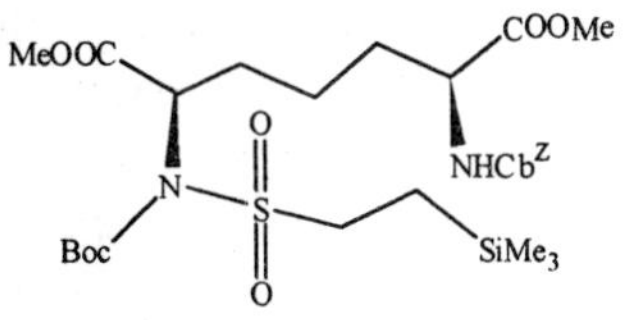

The reaction involves following steps:

1 Formation of intermediate compound **44** by the reaction of dialkylazodicarboxylate (DAAD) and trialkyl or triaryl phosphine.

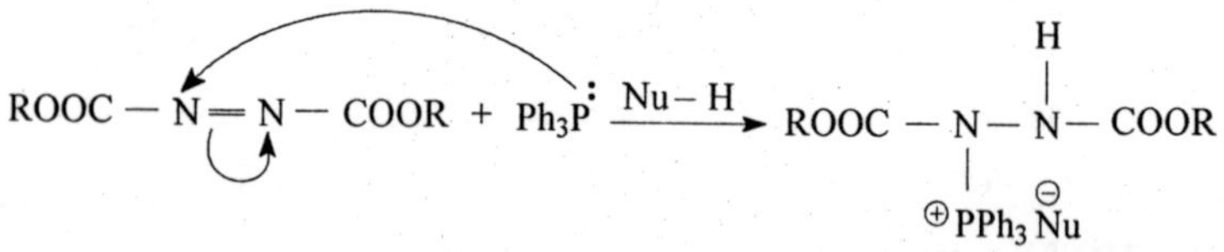

44

2) Condensation of alcohol with **44** takes place to form substituted product **46**. The reaction proceeds with inversion of configuration via the intermediate oxyphosphonium salt **45**.

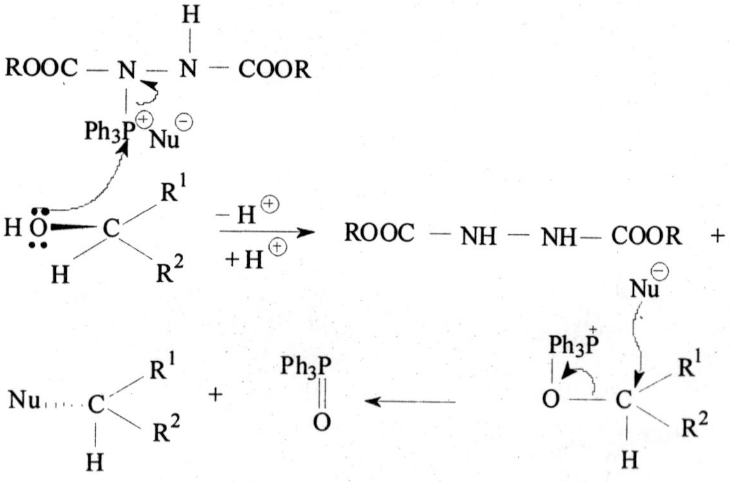

46 **45**

NuH: β-ketoesters $(R-\overset{O}{\overset{\|}{C}}-CH_2-COOR)$, thiols (RSH), Hetrocycles, hydroxymates, oximes ($R_2C=NOH$), imides, phenols (PhOH), carboxylic acid (RCOOH), phosphoric mono ester, phosphoric diester.

11.11. Paterno Buchi reaction

The reaction involves a convenient method for synthesizing oxetanes (four membered cyclic ether). The formation of oxetanes takes place by cycloaddition of carbonyl compounds in the presence of UV light.[32]

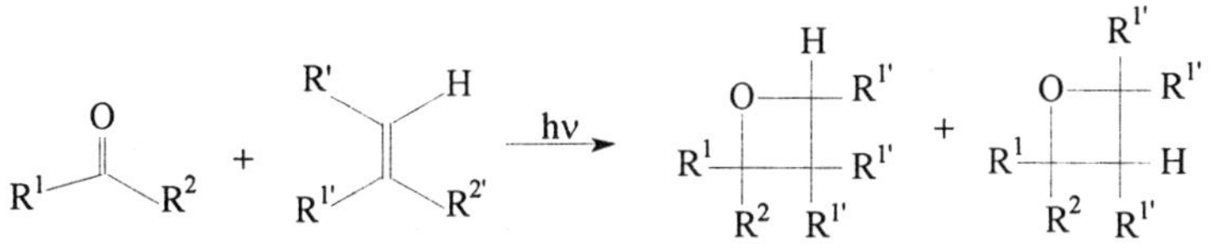

The mechanism of the reaction in presented below.

1. In the presence of ultraviolet rays carbonyl group is converted to its excited triplet state.

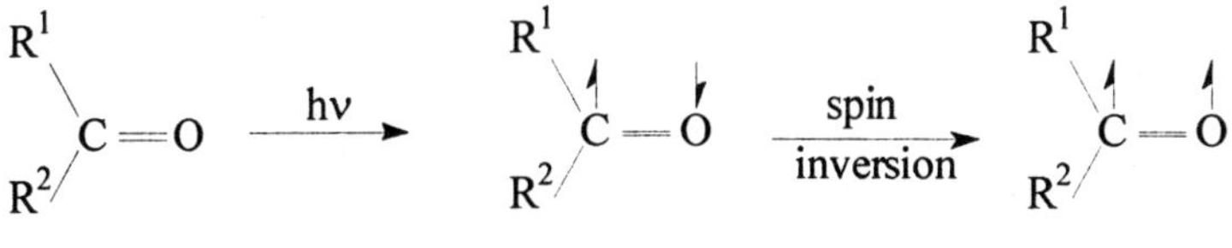

2. The more stable (triplet state) form of carbonyl compound adds to olefin to form diradical.

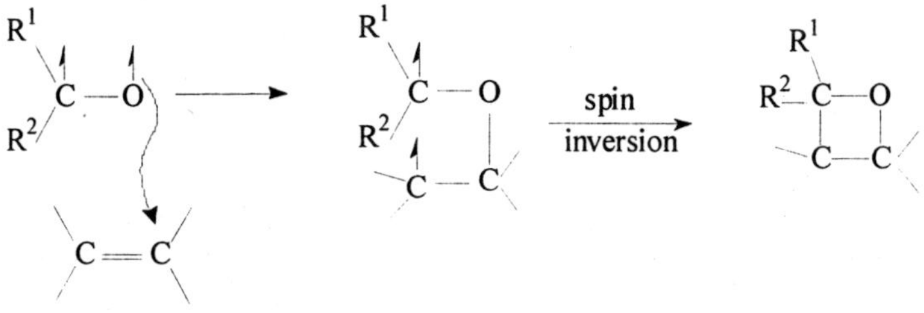

11.12. Simmons Smith reaction

The formation of cyclopropanes by the reaction of olefins with methylene iodide in the presence of Zinc-copper couple is called as Simmons-Smith Reaction. The reaction is highly stereospecific and products is obtained quantitatively.[33,34]

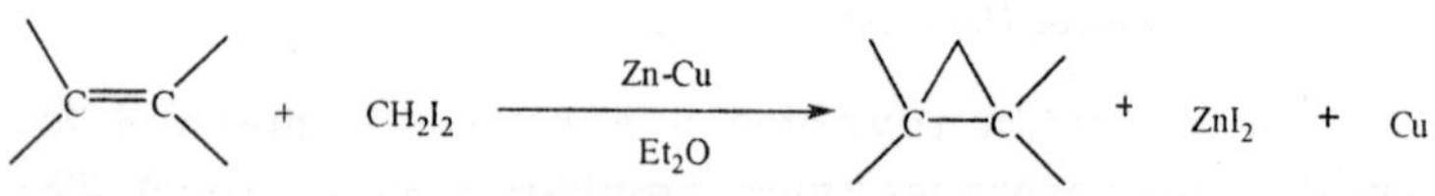

Similarly, cyclopropane derivatives **48** has been synthesized in quantitative yield (93%) from olefin **47** employing CH_2I_2 and diethyl zinc at 25°C in ether as solvent.[35]

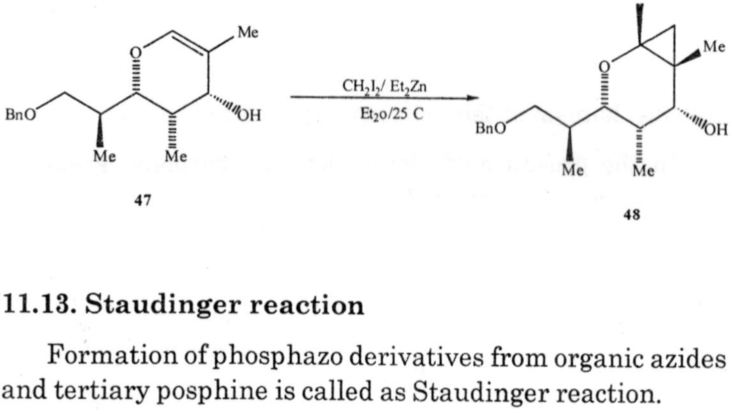

11.13. Staudinger reaction

Formation of phosphazo derivatives from organic azides and tertiary posphine is called as Staudinger reaction.

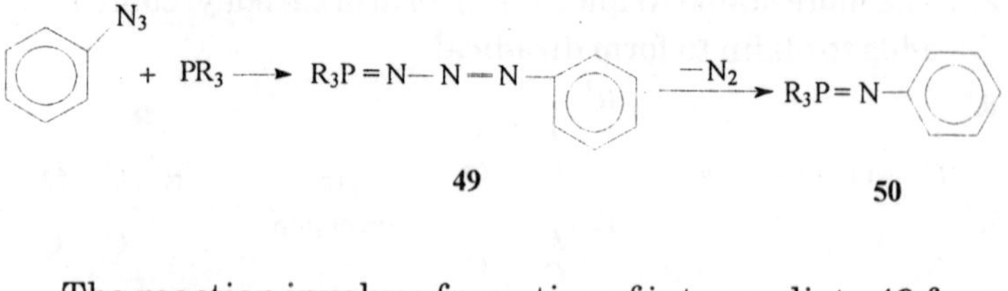

The reaction involves formation of intermediate **49** from the reaction of organic azide with trialkyl / aryl phosphine.

The intermediate **49** then eliminates a molecule of nitrogen to give phhosphazo compound **50**.[36]

11.14. Stobbe condensation

The reaction is used for the synthesis of polycyclic compounds. The reaction involves condensation of carbanion of diethyl succinate with ketone. The reaction takes place in the presence of bases such as potassium tertiary butoxide, sodium hydride etc. The product which is either a cyclic lactone or an alkene or mixture of both is obtained in quantitative yield.

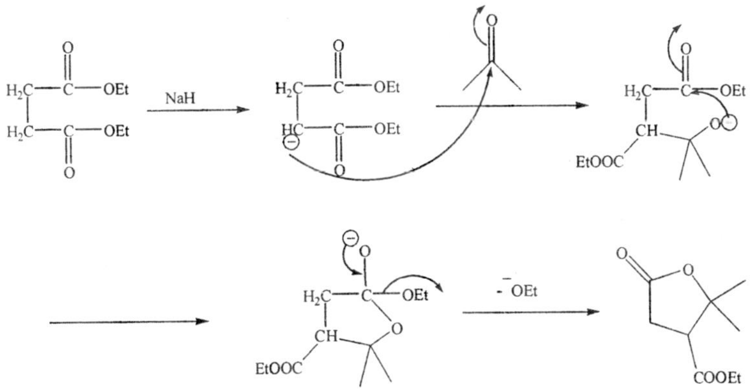

The reaction mechanism leading to the formation of alkene **51** involves following steps.

Formation of carbanion from diethylsuccinate in the presence of base takes place and that adds to ketone to form cyclic adduct **52**. The adduct **52** is finally converted to alkene in the presence of base.

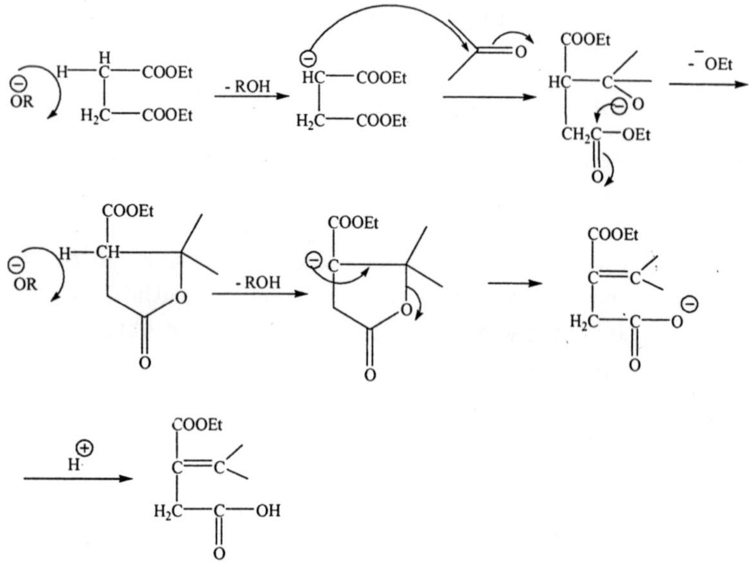

Thus α tetralone **53** has been converted to alkene **54** in the presence of potassium tertiary butoxide.

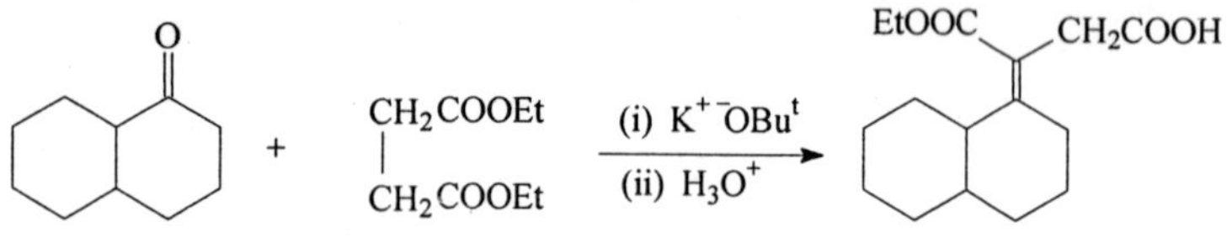

53 **54**

11.15. Ziegler-Natta polymerization

The reaction is about polymerization of monomers containing vinyl group. The polymerization takes place in the presence of Lewis Acid. Thus, polymerization of vinyl monomers in the presence of Lewis acid gives stereo regulated or tactic polymer.[37]

11.16.Ugi reaction

The reaction involves addition of an iminium ion and the conjugate base of a carboxylic acid to an isocyanide. The intermediate, α-adduct, undergoes rearrangement to give α-amino carboxamide derivative **55**.[38]

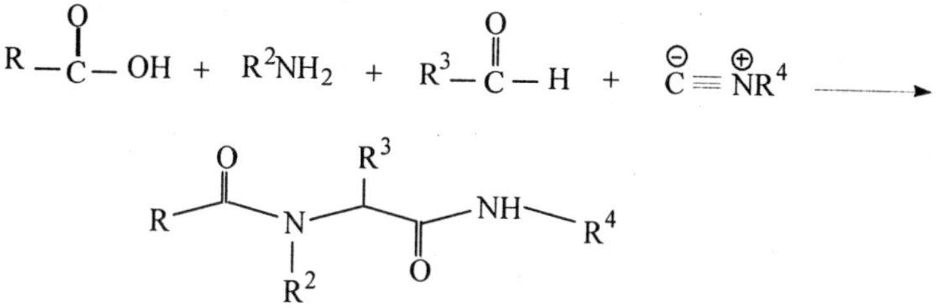

55

The mechanism of the reaction is summarized below:

Formation of imine **58** by the reaction of aldehyde **56** and amine **57**.

1. 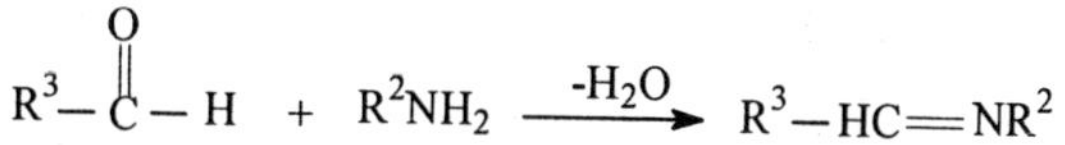

56 **57** **58**

Dissociation of carboxcylic acid into its conjugate base **59**.

2. 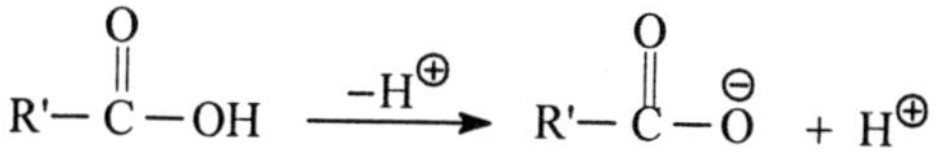

40

Addition of isocyanide to imine **58** to form **60**.

3. 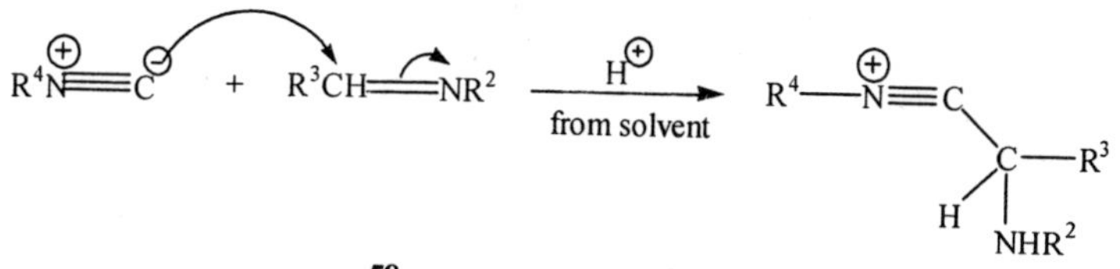

58 **60**

4. Addition of conjugate base of the carboxylic acid **59** to the intermediate 60 to form **61** which undergoes rearrangement involving shift of acyl group to nitrogen to form the product **55**.

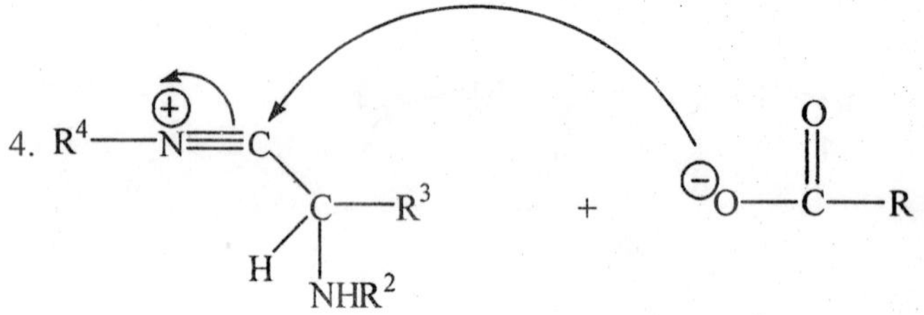

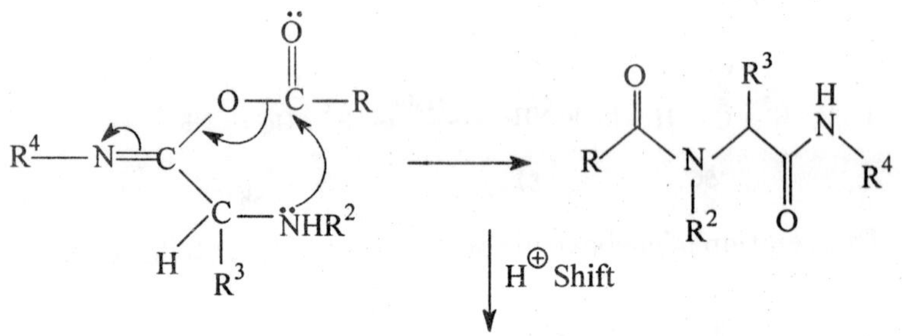

61

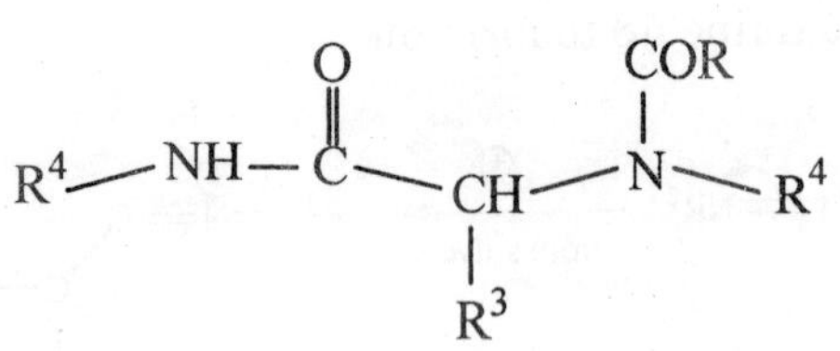

55

Amide **63** has been obtained quantitatively from carboxylic acid **62** through Ugi reaction at low temperature.[39]

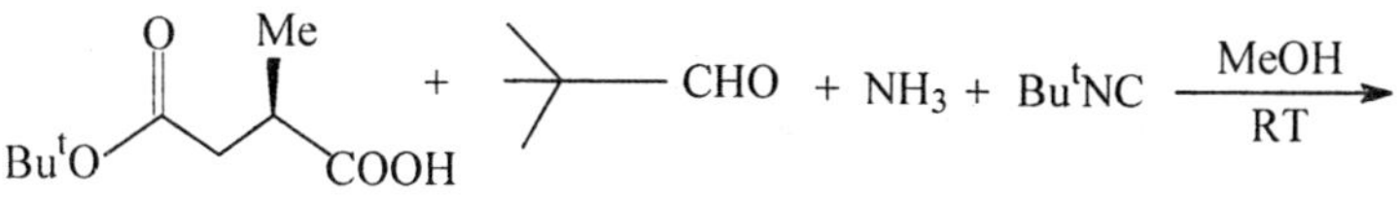

43

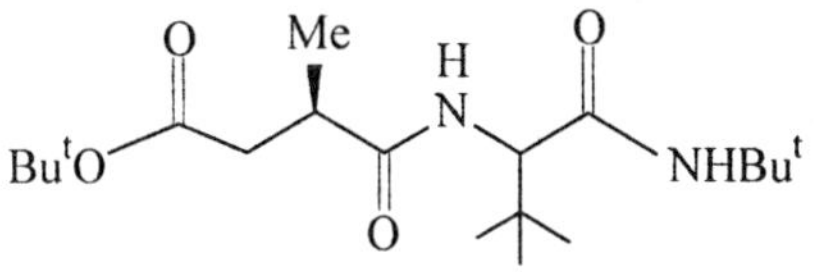

44

11.17. Ullmann reaction

The reaction is used for the synthesis of biaryls. When arylhalides are heated with copper podwer / biaryls are produced in good yield.

$$ArX + Ph - X \xrightarrow{Cu/\Delta} Ph - Ar + CuX_2$$

The order of reactivity of aryl halide is ArI > ArBr > ArCl.

The reactivity of aromatic ring towards coupling is increased when it is substituted with electron withdrawing groups. Further, formation of side products is minimized when one arylhalide is reactive and the other is less reactive.

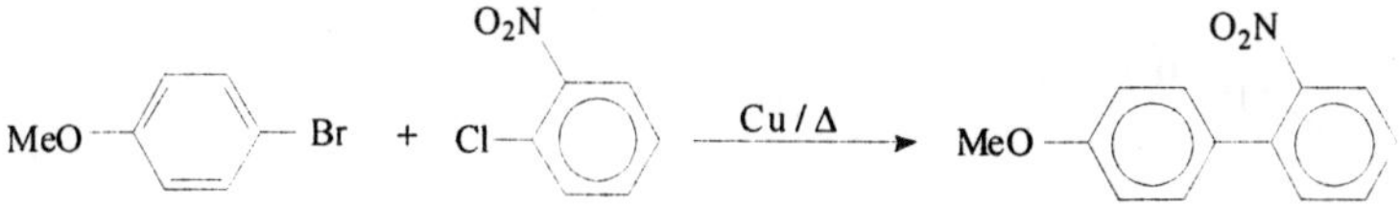

Perhaps the reaction proceeds via free radical pathway.[40]

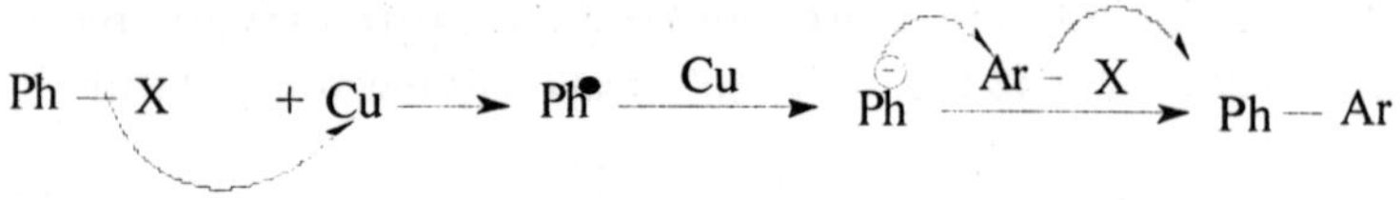

11.18. Wolf Kishner reduction

Conversion of carbonyl compounds to hydrocarbons under basic conditions and in the presence of hydrazine hydrate in a high boiling solvent is called as Wolf Kishner reduction. Ethylene glycol, dimethyl sulfoxida etc.[41,42] may be used as solvents for conducting the reaction.

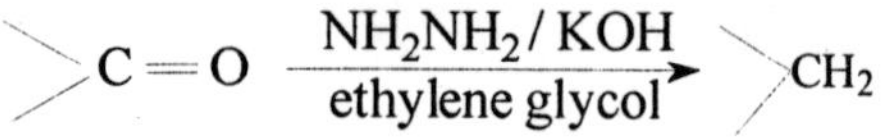

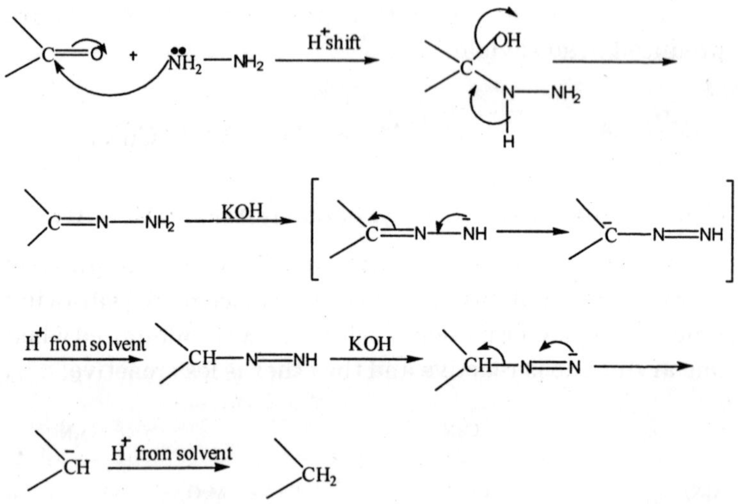

11.19. Wohl degradation

The reaction is used for the conversion of an aldose into

an aldose with one less carbon atom by the reversal of the cyanohydrin synthesis. During the reaction nitrile group is eliminated by treatment with ammoniacal silver oxide.[43]

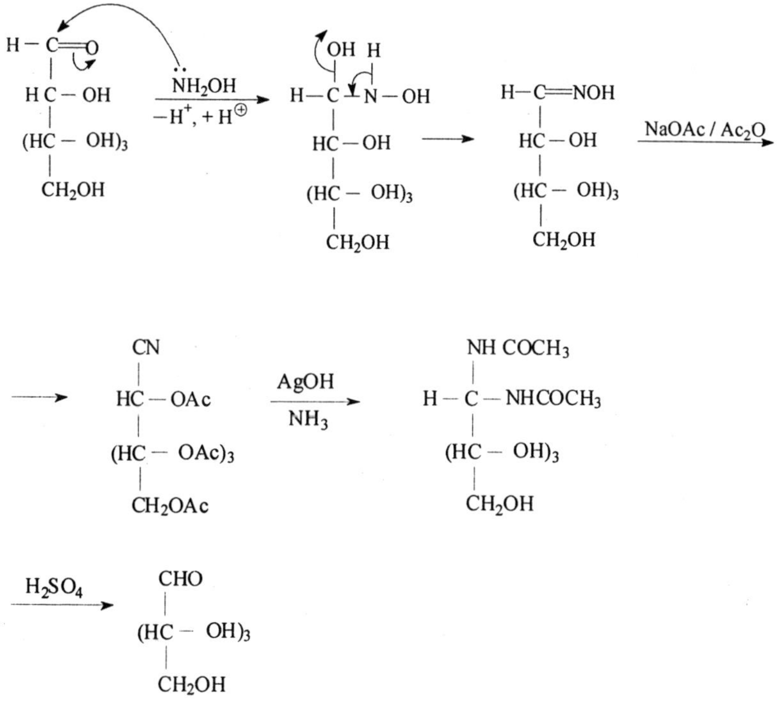

REFERENCES

1. W. N. Lau, K. L. Yeung and R. Martin-Aranda, *Microporous and mesoporous materials*, 2008, **115**, 156-163.

2. R. Khan, M. I.uddin, M. S. Alam, M. M. Hossain and M. R. Islam, *Bangladesh J. Pharmacol.*, 2008, **3**, 27-35.

3. N. S. Rao, B. Baja, Y. K. Srivastava and R. Kumar, *e-Journal Chem*, 2008, **5**, 39-42.

4. S. Balalaie, M. Soleiman-Beigi and F. Rominger, *J Iranian Chem Soc*, 2005, **2**, 319-329.

5. V.M. Markhele, S. A. Sadaphal and M. S. Shingare, *Bull. Catalysis Soc. India*, 2007, **6**, 125-131.
6. S. Pal, J. Mareddy and N. S. Devi , *J Brazilian Chem. Soc.*, **19**, 2008, DOI, 10.1590/so103-50532008 000. 600023
7. R. Lenin and R. M. Raju, *Arkivoc*, XIII, 204-209, 2007.
8. Z. Y. Song, D. R. Bai, H. T. Yu, L. L. Xu, W. X. Zhang, J. B. Meng and T. Matsuura, *Chinese Chemical Lett.* , 2004, **15**, 127-130.
9. J. A. Maclaren, *Australian J Chem*. 1978, **31** (8), 1865-1868.
10. O. Isler et. al., *Helv. Chim Acta* ,1956, **39**, 259,.
11. H. Heusser et. al. *Helv. Chem. Acta.* 1950, **33**, 370.
12. J. F. Arens, *Advanced Org. Chem.* , 1960, **2**, 117-212,.
13. R. Sarger, S. W. Goldstein and W. M. Weieh *J. Med. Chem.*, 1990, **33**, 1859.
14. C. Montagne and M. Shipman, *Synlett*, 2006, 2203.
15. E. Clemmensen, *Ber.*, 46, 1837, 1913; 47, 51, 681, 1914.
16. C. Friedel and J. M. Craft, *Compt. Rend.*, 1877, **84**, 1450.
17. C. C. Price, *Org. React.*, 1946, **3**, 1.
18. P. Gore, *Chem. Rev*, 1955, **55**, 229.
19. M. L. Lewis and A. de Meijere, *Syn Lett*, 1997, 261.
20. B. Helferich and E. Schrnity- Hillebrecht, *Ber*, 1993, **60**, 378.
21. L. Henry, *Compt.Rend.*, 1895, **120**, 1265.
22. H. B. Hass, A. G. Susie and R. L. Heilder, *J. Org. Chem.* 1950, **15**, 8.
23. R. S. Varma, R. Dahiya, S. Kumar, *Tetrahedron Lett*, 1997, **38**, 5131.
24. O. Hinsberg, *Ber.*, 1910, **43**, 901.
25. J. L. Luche, *J. Am. Chem. Soc.* 1978, **100**, 2226
26. D. Turner and P. Vogel, *Synlett*, 1998, 304

27. D. J. Hart, J. Li, W-L, Wu and A. P. Kozinkowski, *J. Org. Chem.*, 1997, **62**, 5023.

28. O. Mitsunobu and M. Yamada, *Bull. Chem. Soc. Jpn,* 1967, **40**, 2380.

29. O. Mitsunobu, *Synthesis,* 1981, 1

30. T. Bach, K. Kather and O. Kramer, *J. Org. Chem.* 1998, **63**, 1910.

31. Y. Gao, P. Lane- Bell and C.J. Vederas, *J. Org. Chem.* 1998. **63**, 2133.

32. E. Paterno,andG. Chieffi, *Gazz. Chim. et al.*, 1909,**39**; Buchi et al., *J. Am. Chem. Soc.* 1954, **76,** 4327.

33. H.E. Simmons and R.D. Smith, *J. Am. Chem. Soc.,* 1958, **80**, 5323, *ibid*, 1959, 81, 4256.

34. J.M. Denis, J.M. Girard and J.M. Conia, *Synthesis,* 1972, 549.

35. P. Bertinato, E. J. Sorensen, D. Meng and S. J. Danishefsky, *J. Org. Chem,* 1996, **61**, 8000.

36. H. Staudinger and J. Meyer, *Helv. Chem. Acta,* 1919, **2**, 635.

37. K. Leiglar et al., *Angew. Chem.*, 1995, 67, 426, 541; G. Natta; *Ibid*, 1956, **68,** 393.

38. I. Ugi, *Angew. Chem.* Int. Ed., Engl. 1962, **1, 8**.

39. C.D. Floyd, L.A. Harnett, A. Miller, S. Patel, L. Saraglou and M. Whitlaker *Synlett,* 1998, 637.

40. F. Ullmann, *Ann.* 1904, **332**, 38.

41. N. Kishner, *J. Russ. Phys. Chem. Soc.,* 1911, 43, 582; L. Wolf, *Ann.*, 1912, **394, 86**.

42. Huang Minlon, *J. Am. Chem. Soc.,* 1946, **68**, 2487.

43. A. Wohl, *Ber,* 1893, **26**, 730; G. Zemplen, *Ber*, 1926 **59,** 1254.

44. A. Hantzch, *Ber*, 1885, **18**, 1744.

12

Molecular Rearrangement

Migration of an atom or group of atoms from one atom to another within the molecule or from one molecule to another is called as molecular rearrangement. It is shown by a curved arrow (—o→). There are two types of rearrangements:

Inter-molecular rearrangements

During the rearrangement when the migratory group detaches itself from the molecule and becomes attached to other reactive part of the molecule, it is called as intermolecular rearrangement.

Intramolecular rearrangement

However, when migratory group does not leave the molecule during the migration it is called as intramolecular rearrangement

There are various kinds of rearrangements depending upon reactive site of the molecule to which migratory groups becomes attached. Thus, a group may migrate to an electron deficient carbon, nitrogen, oxygen or an electron rich carbon. (Fig 1).

In the rearrangement involving electron deficient site, the migrating group migrates along with its bonding electron pair. Most commonly, shift of group takes place from one carbon atom to the adjacent carbon atom (1,2-shift).

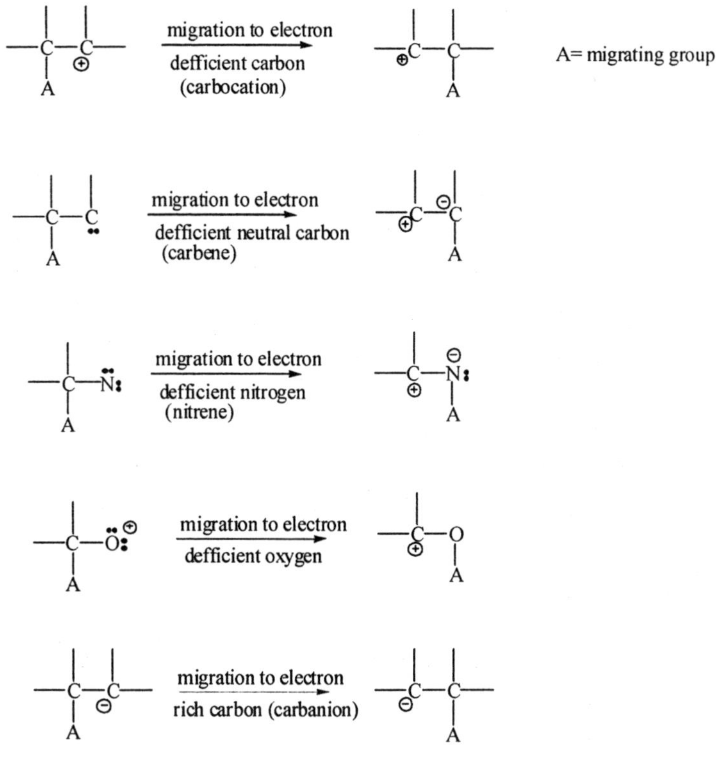

Fig 1 Rearrangement of electron deffficient and electron rich systems.

Various kinds of rearrangements are discussed below:

12.1. Rearrangements involving migration to electron deficient carbon

When a carbocation is formed during a reaction either through ionization or protonation of a double bond, then it may interact with a nucleophile to form a substituted product or may remove a proton from unsaturated product or it may form a new carbocation through rearrangement.

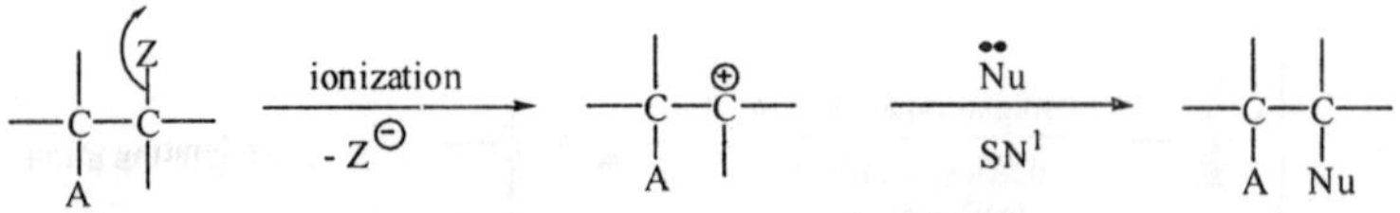

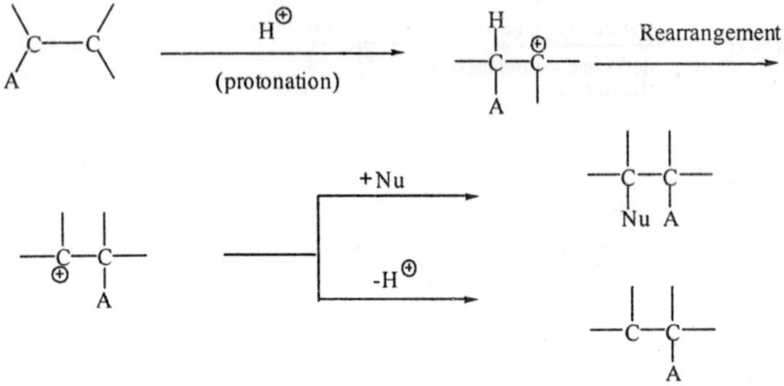

Thus, e.g. when n-butanol is heated in the presence of some Lewis acid, then rearranged 2-butene is formed as major compound rather than 1-butene.

a. 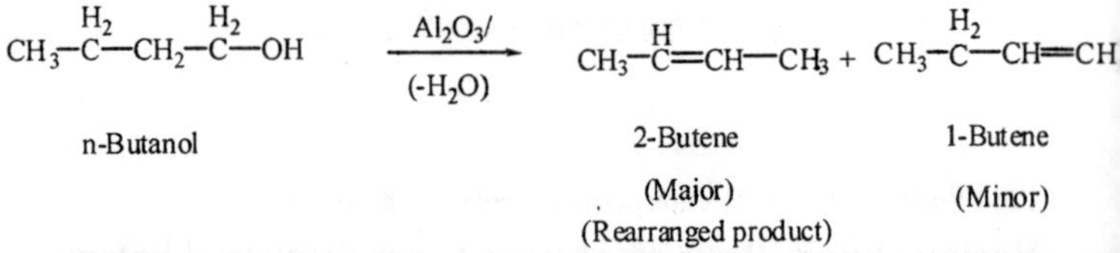

Wagner-Meerwein rearrangement

This involves rearrangement of alcohols under acidic conditions. These rearrangements involve carbocation as intermediate. A number of these rearrangements have been used synthetically for ring enlargement and ring contraction.

Thus, amino cyclohexanol **1** rearranges to aldehyde **2** in the presence of nitrous acid.

b. 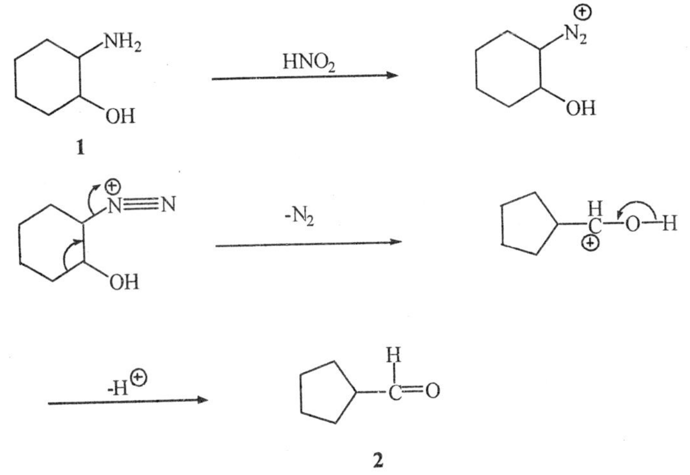

Similarly amine **3** rearranges to cyclobutanol **4**.

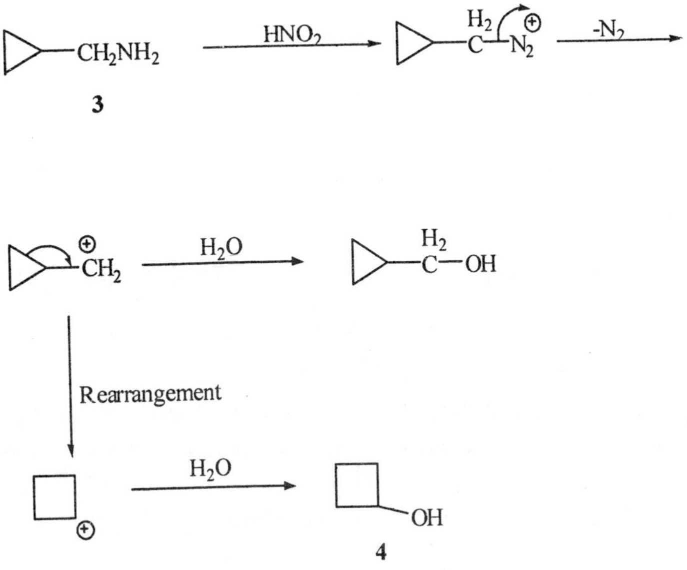

Similarly, iodocyclohexanone **5** rearranges to cyclopentyl nitromethane **6** upon treatment with silver nitrite.

c. 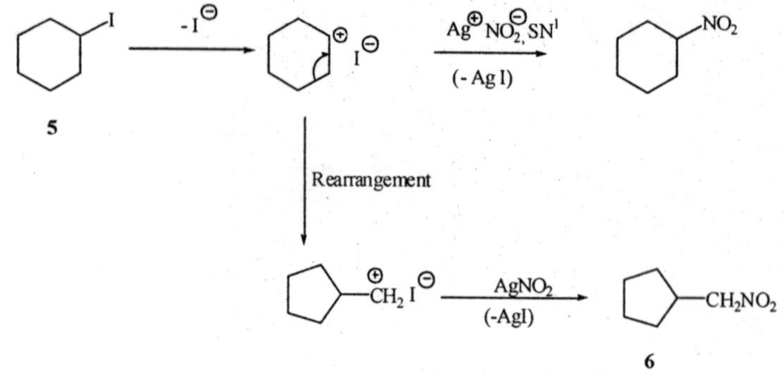

Thus, Wagner-Meerwein rearrangement results in change in the carbon skeleton involving carbocation as intermediate.

Regarding the mechanism of Wagner-Meerwein rearrangement, the first step is the formation of carbocation which may take place in several ways e.g.

(i) through loss of water, as in example, (a) from protonated alcohol.

(ii) Through loss of nitrogen from aliphatic diazonium salt (example b)

(iii) Through loss of halide ion from alkyl halides (example c)

(iv) Through addition of electrophile to a double bond following markownikoff's rule.

In the second step, the migration of atom or group of atoms to adjacent carbocation takes place to form rearranged carbocation which may combine with a nucleophile or may lose a proton to give unsaturated compound (Fig 2).

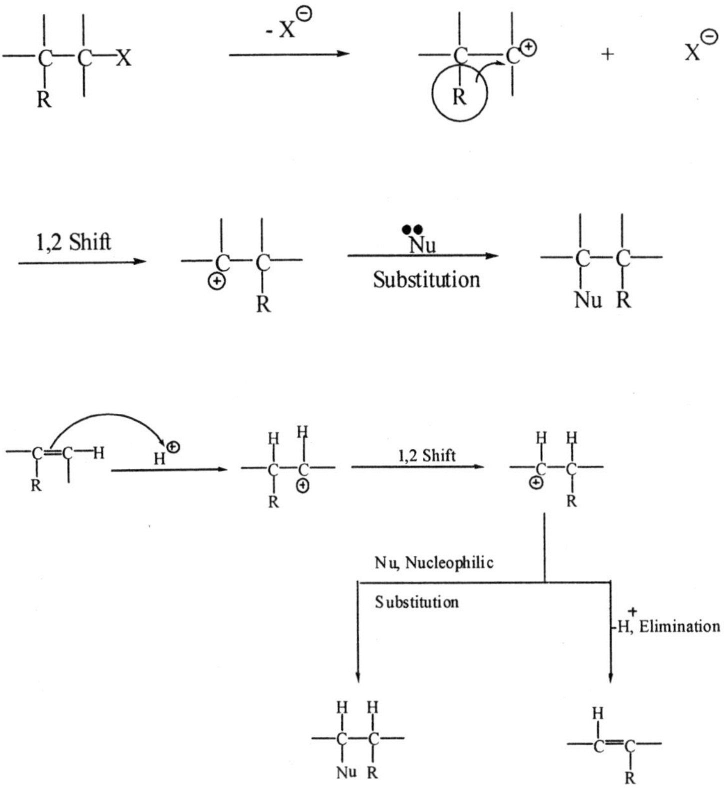

Fig 2 Mechanism of rearrangement involving migration to electron deficient carbon.

Direction and stereochemistry of rearrangement

Aptitude of migration

During migration of a group there exists a choice as to which group migrates towards the electron deficient atom. The migrating group forms a three membered cyclic electron deficient transition state. Thus, group which stabilizes carbocation lowers the energy barrier and increases the rate of migration, hence rate of the reaction. The order of preference of various migratory groups is :

Aryl > tertiary alkyl > secondary alkyl > primary alkyl> CH_3 >H

The aryl group distributes the charge through its π electrons as shown below:

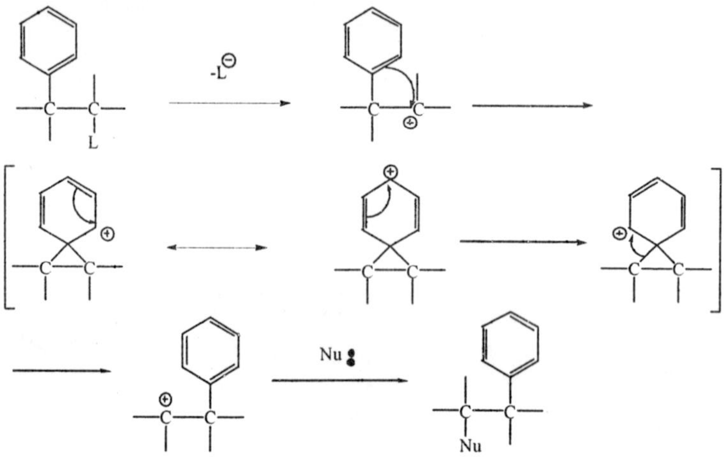

The plane of the aromatic ring is at right angle to the three carbon ring. The rearrangement of phenyl group is, thus, an internal electrophilic substitution on the migrating aromatic ring.

Stereochemistry of Wagner- Meerwein rearrangement

Usually Wagner- Meerwein rearrangements are stereospecific. The migrating group approaches the carbocation from back side. This results inversion of configuration at the carbon which was previously linked to the leaving group (Fig. 3.).

During migration the configuration of migrating group is retained.

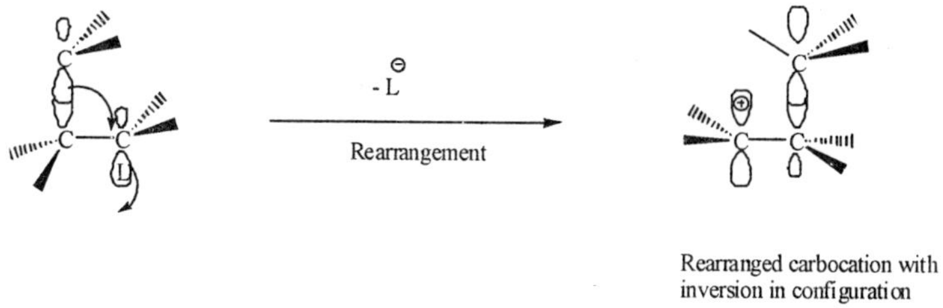

Fig 3 Stereochemistry of rearrangement

Thus, solvolysis of optically active threo-3-phenyl-2-butyltosylate forms recemic threo acetate whereas erythro-3-phenyl-2-butyl tosylate gives optically pure erythro acetate.

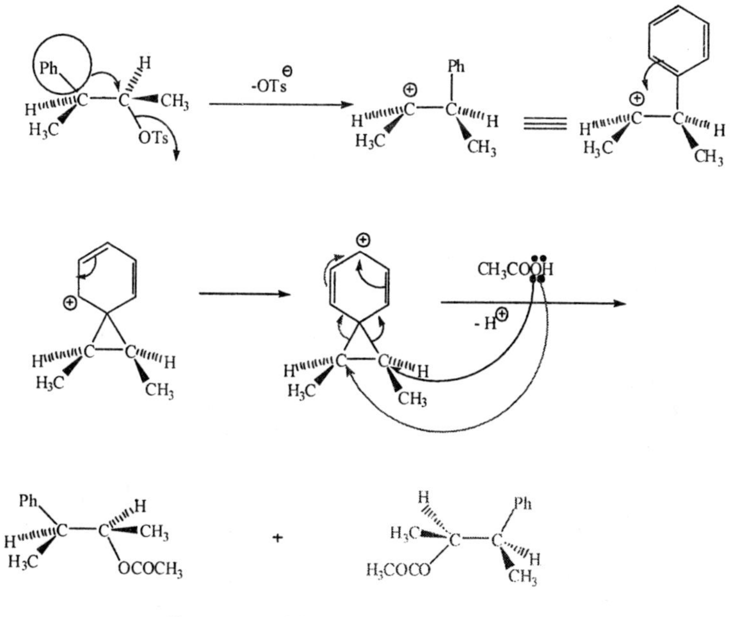

Threoacetate (recimic mixture)

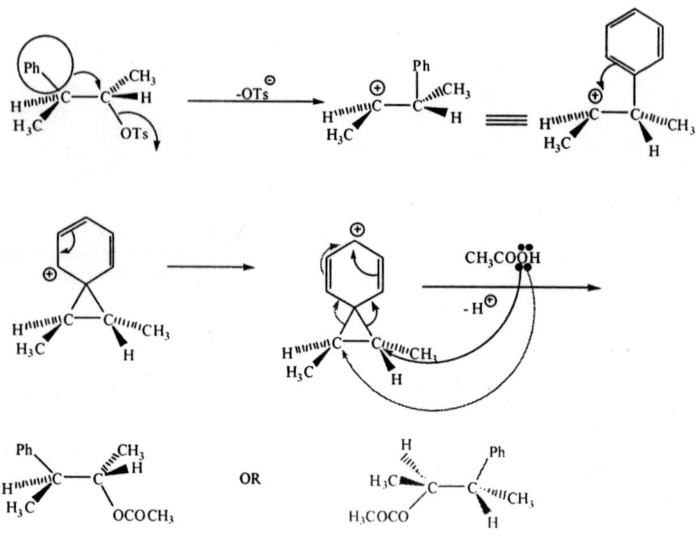

erythreo acetate (optically active)

Some important examples of electron deficient (cationic) skeletal rearrangements.

1. Pinacol-pinacolone rearrangement

1,2-Diols when heated in the presence of a Lewis acid undergoes rearrangement to form aldehydes or ketones.[1]

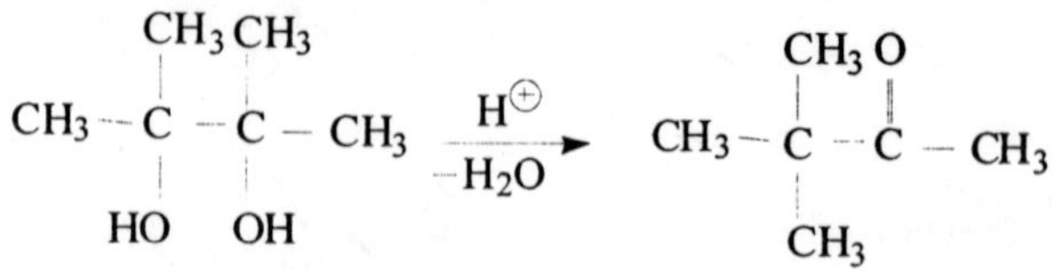

The reaction involves following steps.

1. Protonation of one of the hydroxyl group to form oxonium ion **7**.

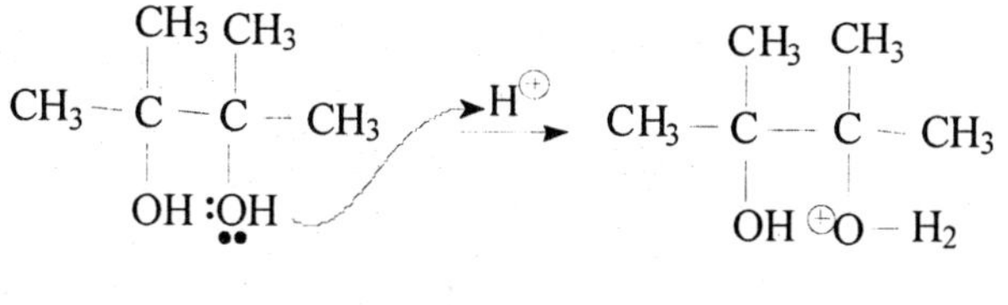

7

2. Loss of water molecule from oxonium ion **7** takes place to form carbocation **8**.

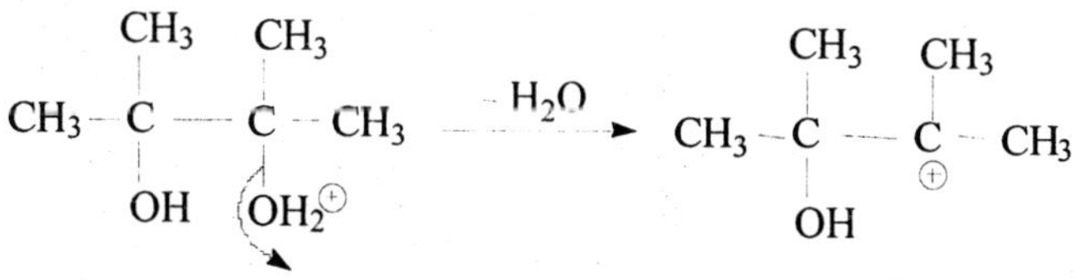

7 8

3. Since carbocations are susceptible to rearrangement, carbocation **8** rearranges to **9** involving 1,2 methylshift (migration of methyl group along with its bonding electron pair to carbonation takes places).

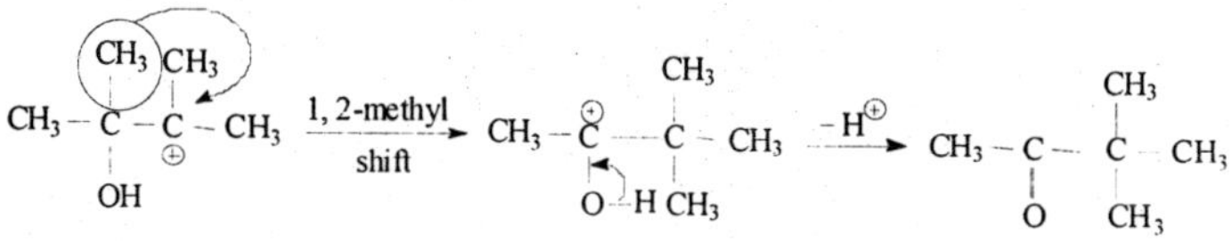

8 9

When diol is unsymmetrical then dehydration takes place in such a way so as to form more stable carbocation. Further the formation of the product depends on the nature of migrating groups. The order of migrating aptitude of different groups is Ph > H > R, e.g.

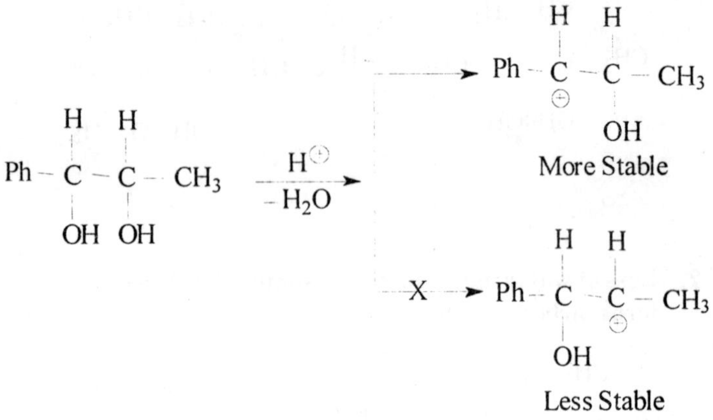

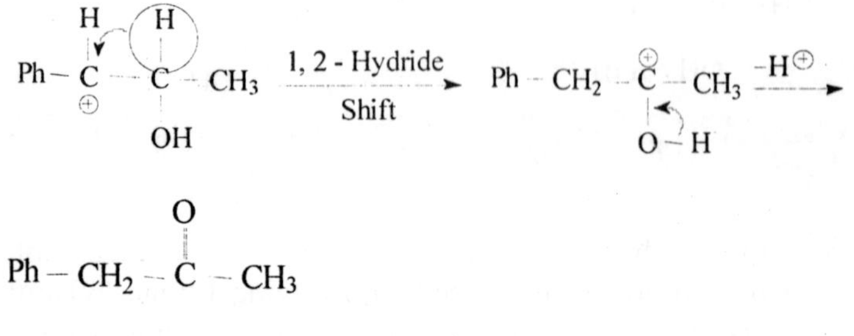

Synthetic utility

1. Spiro ketone **11** is obtained from diol **10** in the presence of an acid.[2]

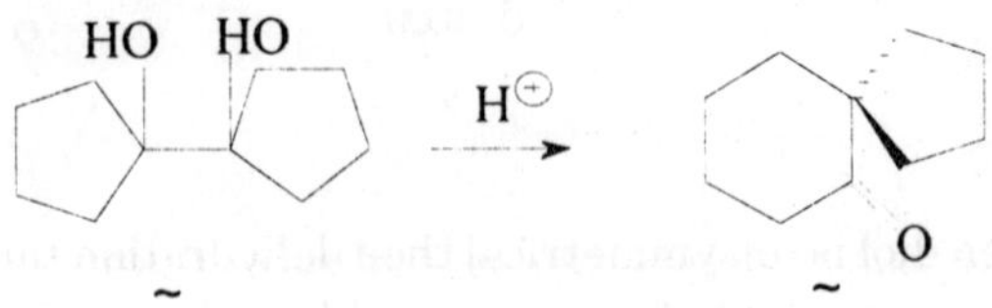

2. β-amino alcohol **12** when treated with nitrous acid, rearranged ketone **13** is obtained.

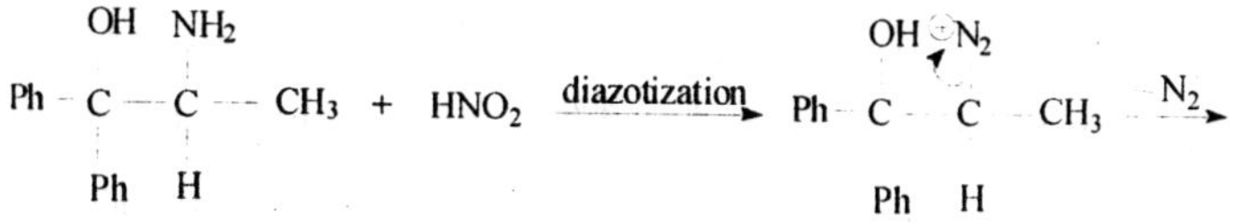

12

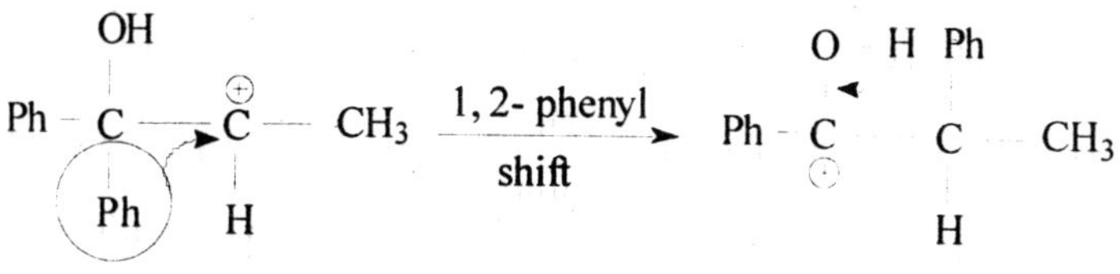

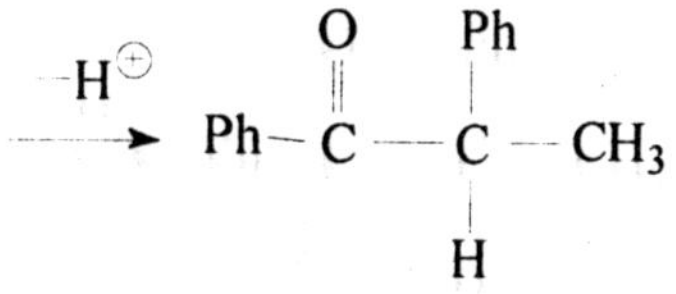

13

3. Halohydrins **14** have been conveniently converted to ketones **15** upon treatment with silver ions.

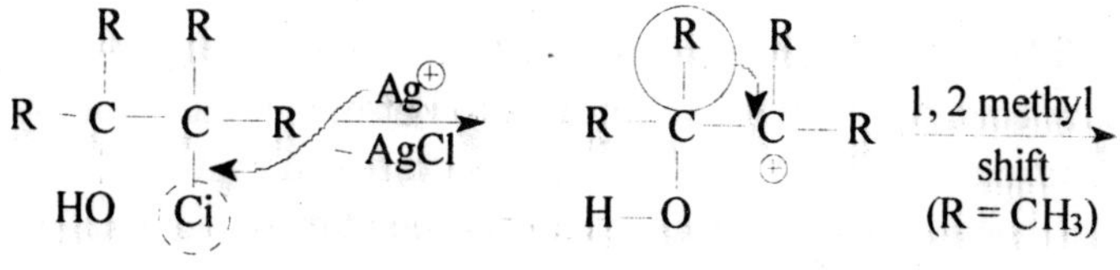

14

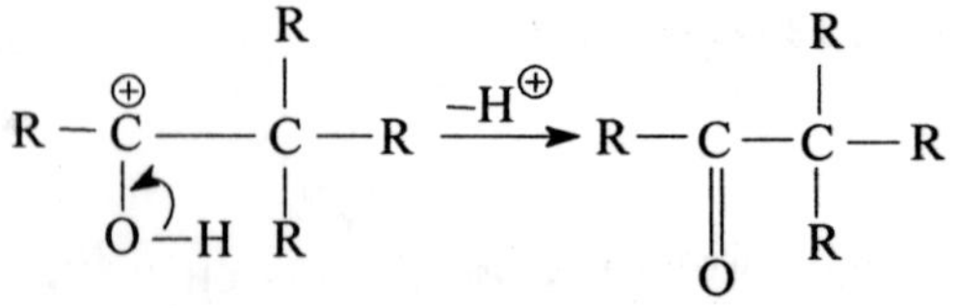

15

4. Epoxide **16** has been transformed to ketone **17** in the presence of a Lewis acid.

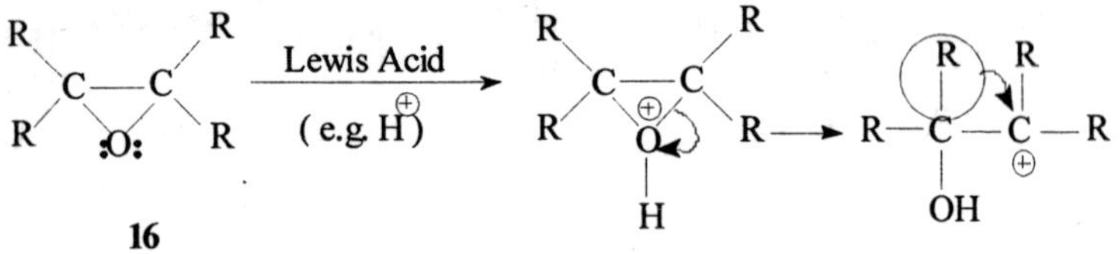

16

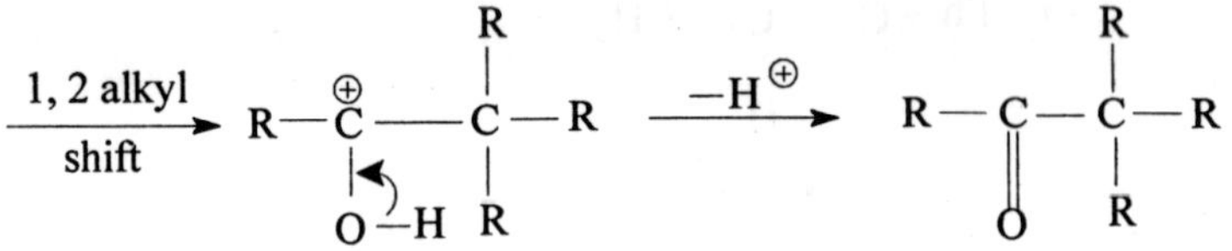

17

2. Wolff rearrangement

Thermal, photochemical or catalyst induced rearrangements of diazoketones to ketenes is called as Wolf rearrangement. The mechanism is observed in Arndt Eistert Synthesis which is an important method for the homologation of carboxylic acids.[3]

Arndt Eistert Synthesis: The reaction consists of following steps:

1. Formation of acid chloride **19** by treatment of carboxylic acid **18** with thionyl chloride or phosphorous pentachloride.

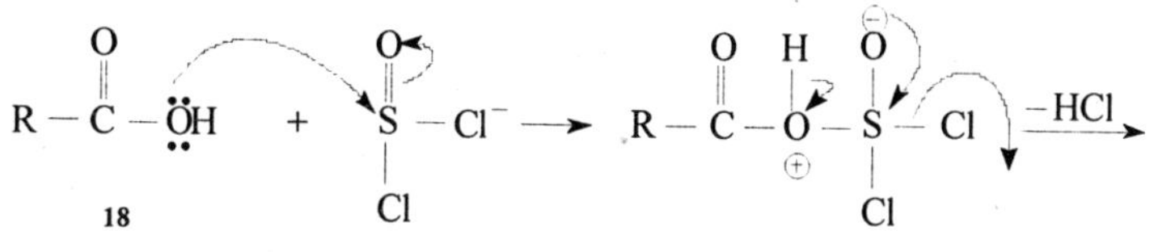

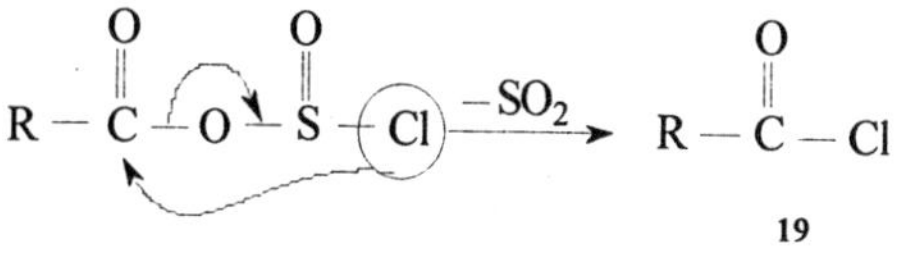

2. Conversion of acid chloride **19** to diazoketone **20** by treatment with diazomethane.

3. Diazoketone **20** undergoes Wolff rearrangement to form intermediate ketene which can be converted to carboxylic acid **21** upon treatment with colloidal silver, ester **22** upon treatment with alcohol, amides **23,24** upon treatment with ammonia or amines.

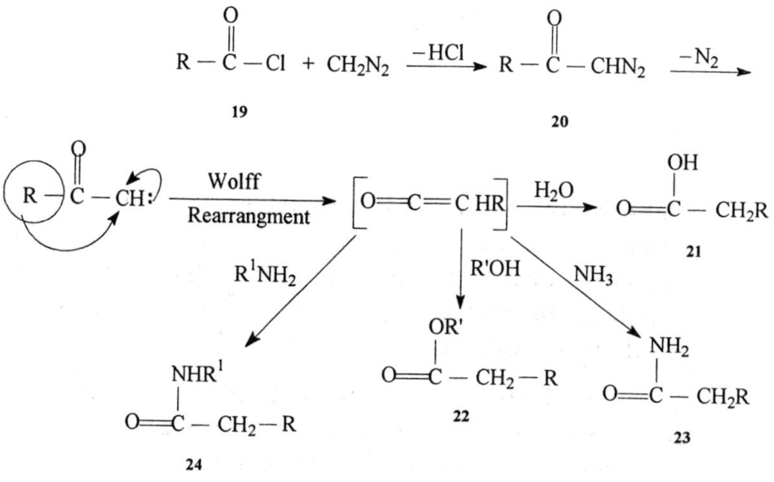

3. Benzil-Benzilic acid rearrangement

The reaction involves rearrangement of benzil **25** (α-diketones) to benzilic acid **26** in the presence of a base.[4] The rearrangement occurs via migration of phenyl group.

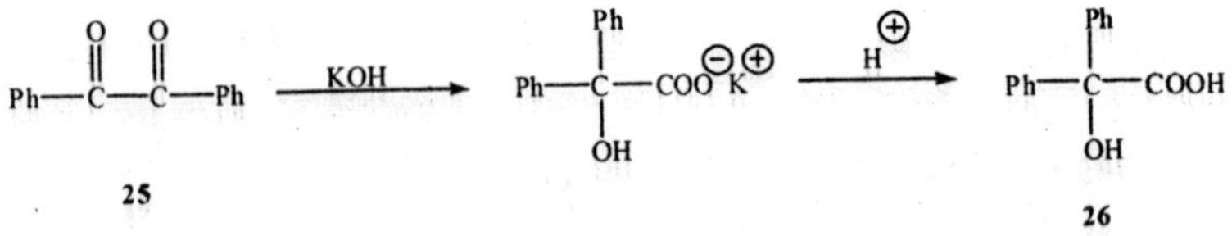

The reaction mechanism is shown below.

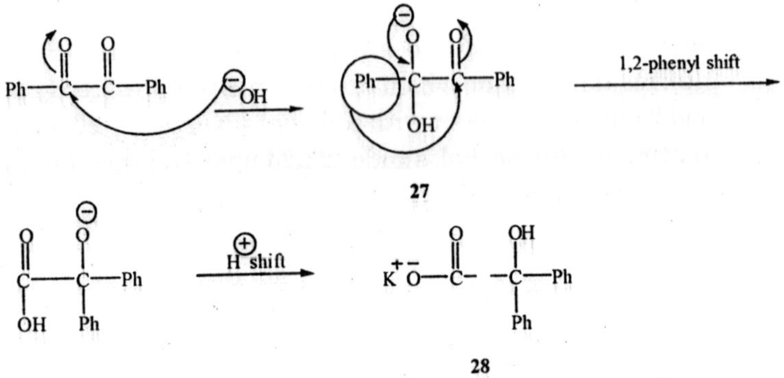

The hydride ion adds to carbonyl group to form intermediate **27** which undergoes rearrangement involving migration of phenyl group along with its bonding electron pair to adjacent carbonyl group followed by proton shift to give salt of benzilic acid **28**.

Free acid **26** is regenerated from its salt **28** upon acidification. The formation of carboxylic acid is feasible as carboxylate ion is stabilized by resonance.

12.2. Rearrangements involving migration to electron defficient nitrogen

These are the rearrangements which involve migration of a group to electron deficient nitrogen known as nitrenes.

The electron deficient nitrogen is generated by loss of a leaving group attached to nitrogen. The rearrangements involving migration of a group to nitrene follows a concerted mechanism, i.e. migration of group to nitrene and removal of leaving group from nitrogen takes place simultaneously. Migration usually involve 1,2 shift and driving force for these types of rearrangements is provided by removal of small neutral molecules such as water, nitrogen etc.

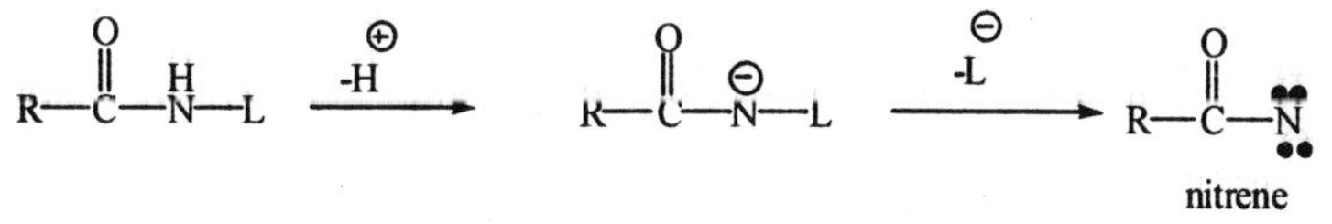

Some important examples involving migration to electron deficient nitrogen are discussed below.

1. Hofmann Rearrangement

The reaction is used for the synthesis of primary amines by treatment of primary carboxylic acid amide with bromine and aqueous or alcoholic potash (or potassium hypobromite).[5] The reaction proceeds via intermediate isothiocyanate **29**.

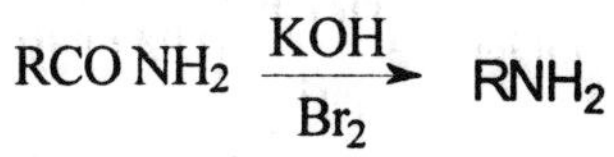

The reaction mechanism involves following steps.

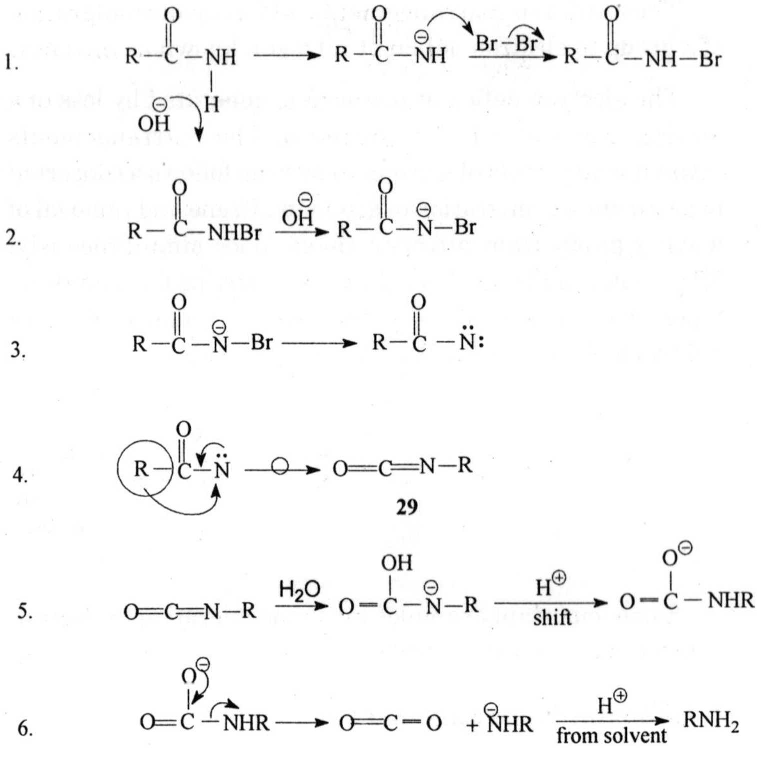

The proposed mechanism is plausible as intermediates.

$$RCONHBr,\quad R-\overset{\overset{\displaystyle O}{\|}}{C}-\overset{\ominus}{N}-Br \quad \text{and} \quad O=C=N-R$$

In steps 1,2, 4 have been isolated.

2. Lossen rearrangement

The reaction is used for the synthesis of primary amines.[6]

The reaction takes place by treatment of hydroxamic acid with some inorganic acids such as hydrochloric acid to form nitrene **30** intermediate. The nitrene **30** intermediate undergoes rearrangement by shift of methyl group to form isocyanate **31** The hydrolysis followed by decarboxylation of isocyanate gives primary amine **32**.

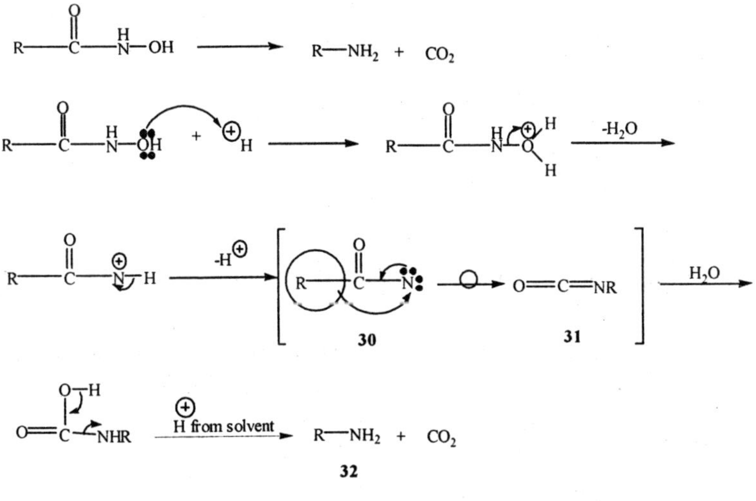

Sometimes hydroxamic acid **33** is converted to acyl derivative of hydroxamic acid **34** which upon base treatment undergoes rearrangement to form amine **35**.

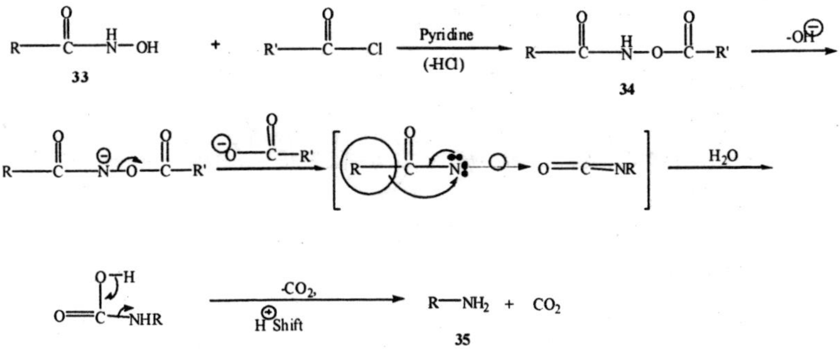

3. Curtius rearrangement

Formation of isocyanates **37** by thermal decomposition of acyl azide **36** is called as curtius rearrangement.[7]

$$\underset{\mathbf{36}}{RCON_3} \xrightarrow{\Delta} \underset{\mathbf{37}}{R-N=C=O}$$

The rearrangement takes place during the synthesis of primary amine from carboxylic acid.

The mechanism involves following steps.

1. Formation of acid chloride **38** of carboxylic acid by treatment with thionyl chloride

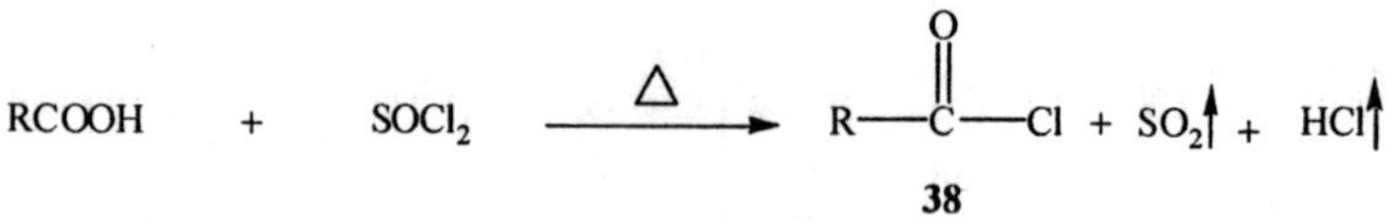

38

2. Conversion of acid chlorides **38** into acyl azide **36** by carrying out reaction with sodium azide.

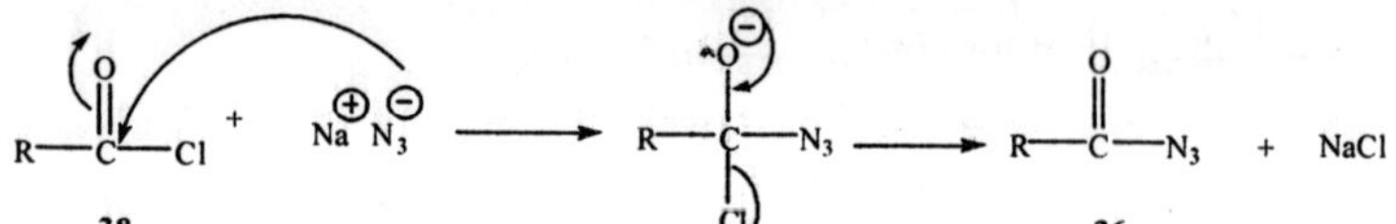

38 **36**

3. Acyl azide **36** is treated in an inert solvent to form isocyanate **37** through curtius rearrangement.

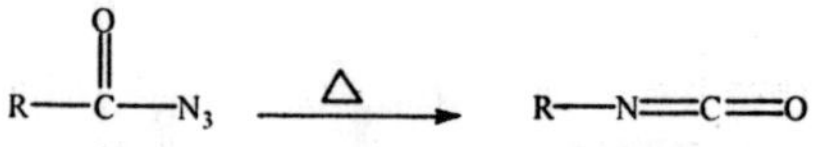

36 **37**

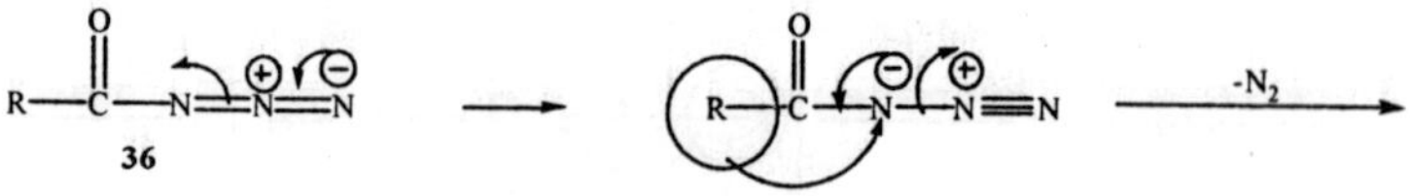

37

3. Hydrolysis of isocynates **37** forms carboxylic acid **39** which decarboxylates to form primary amine **40**.

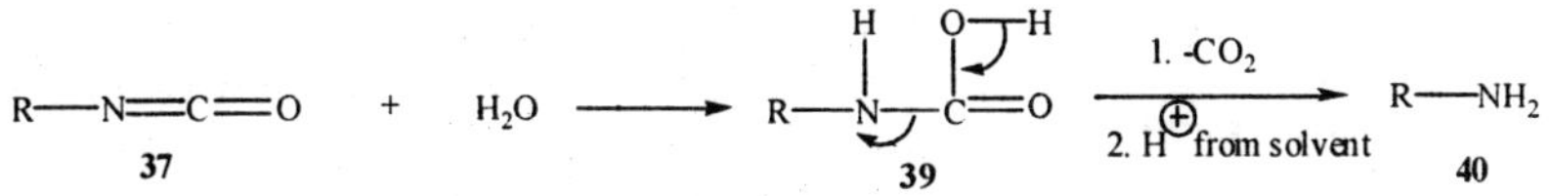

4. Beckmann rearrangement

Conversion of ketoxime **41** to primary amine **43** *via* intermediate amide **42** in the presence of a Lewis acid such as phosphorous pentachloride is called as Beckmann rearrangement.

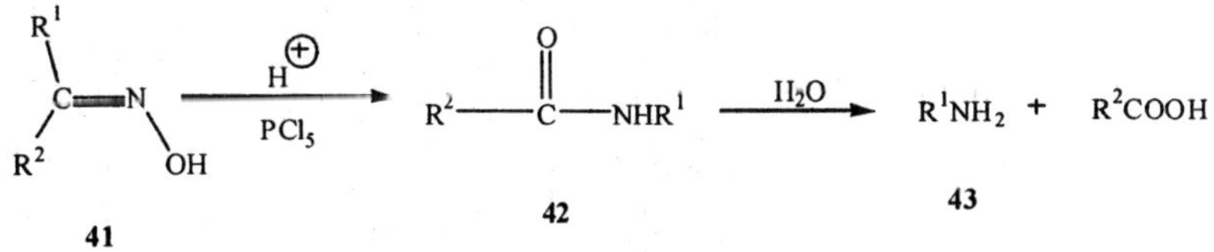

The mechanism of the reactioninvolves following steps.

(1) Protonation of oxime **41** to **44** which undergoes dehydration to form intermediate **45**.

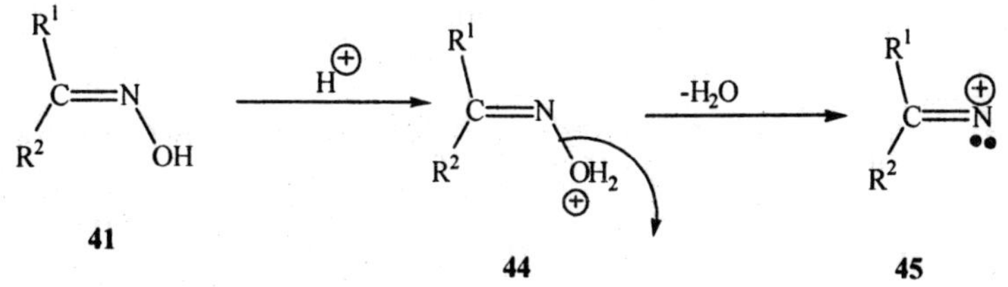

(2) Migration of R^1 (alkyl group which is opposite to OH group in oxime **41**) to positively charged nitrogen takes place to form intermediate **46**.

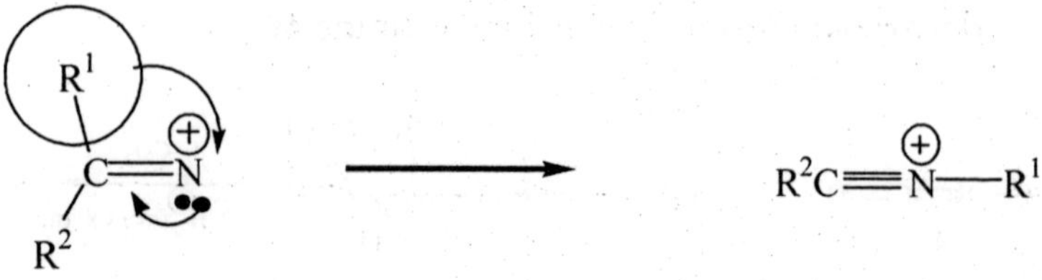

45 **46**

(3) Hydrolysis of intermediate **46** forms amide **42** which is then converted to the final product amine **43**.

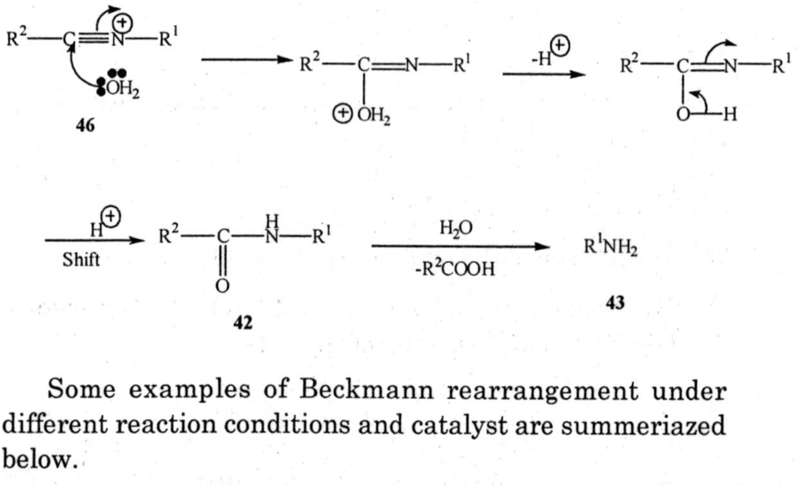

46 **42** **43**

Some examples of Beckmann rearrangement under different reaction conditions and catalyst are summeriazed below.

1. Synthesis of amide **48** from ketoxime **47** in the presence of trichlorotriazine.

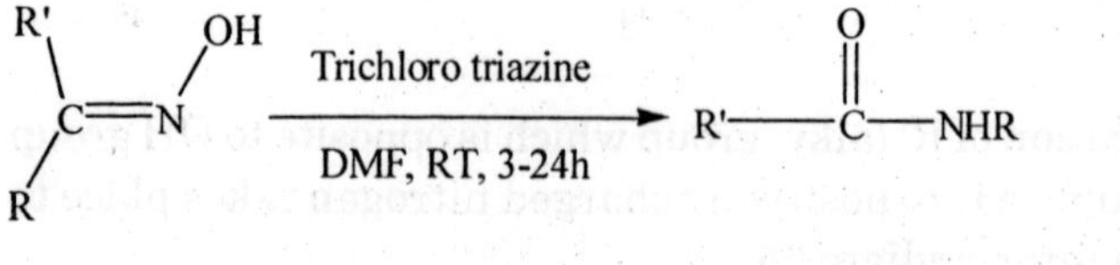

[Ref 8]

47 **48**

2. Amide **50** was synthesized from **49** under the influence of chloral.

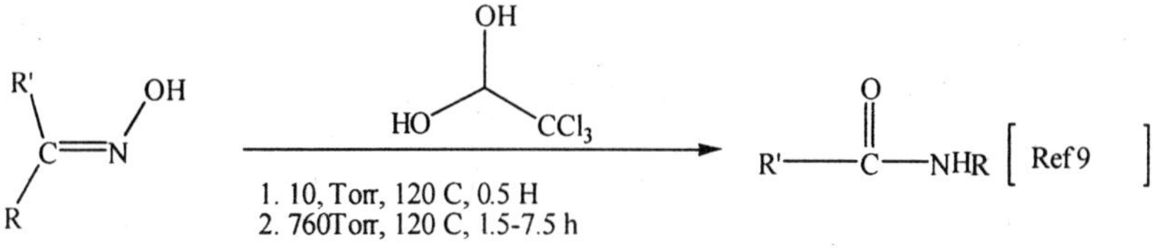

49 **50**

3. Mercury catalysed Beckmann rearrangement of ketoxime.

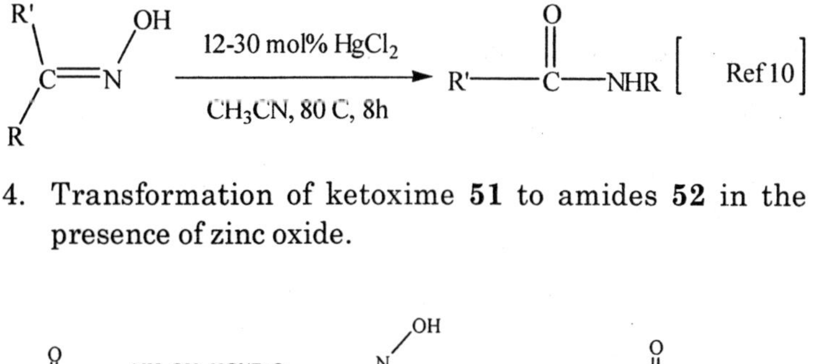

4. Transformation of ketoxime **51** to amides **52** in the presence of zinc oxide.

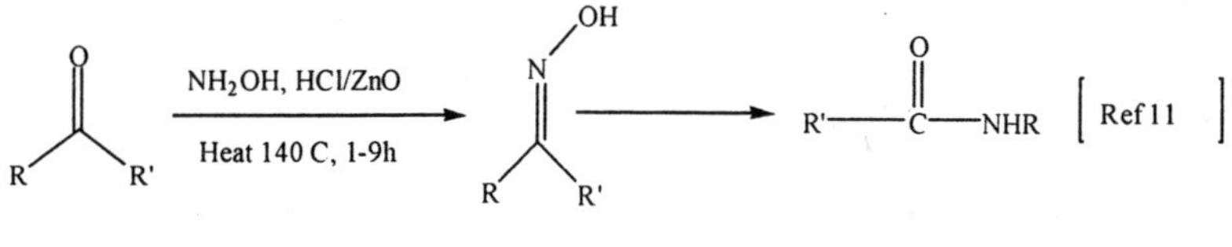

52

51

12.3. Rearrangements involving migration to electron deficient oxygen

Rearrangement involving migration of a group to electron deficient oxygen requires either breaking of O-O bond or cleavage of a bond in which leaving group is an oxyanion.

Some examples of rearrangements involving migration of a group to electron deficient oxygen are as follows:

1. Baeyer-Villiger oxidation

The oxidation of ketones **53a-b** to esters **54** / lactones **55** with peracid is known as Baeyer- villiger oxidation.[12]

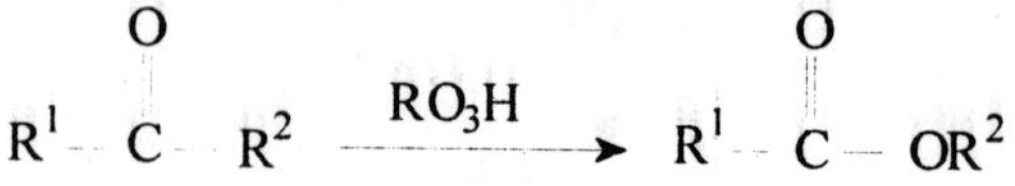

53a **54**

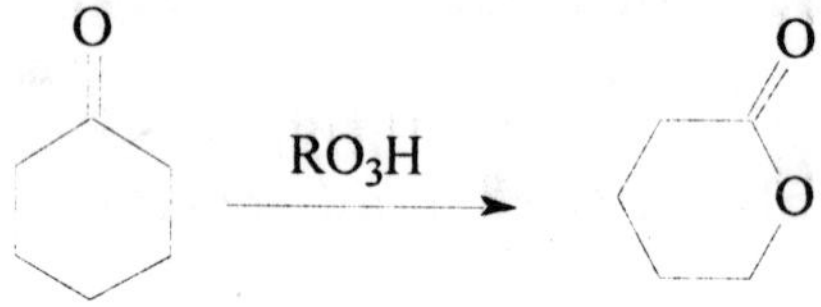

53b **55**

The reaction involves following steps.

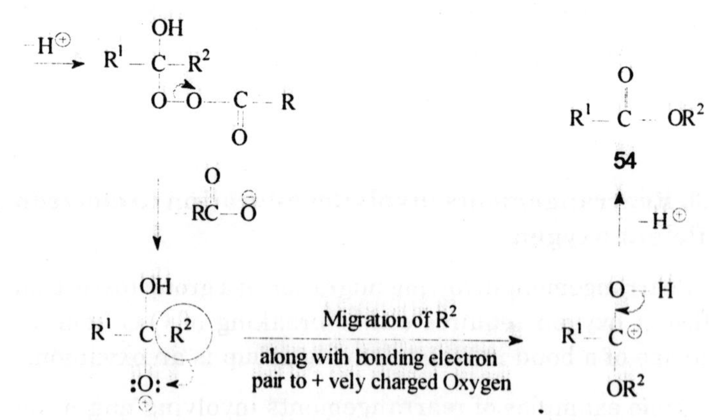

2. Dakin rearrangement

It is a reaction similar to Bayer Villigar Oxidation. In this reaction aldehyde or ketonic group in phenolic aldehydes or ketones is converted to hydroxyl group upon treatment with alkaline hydrogen peroxide.

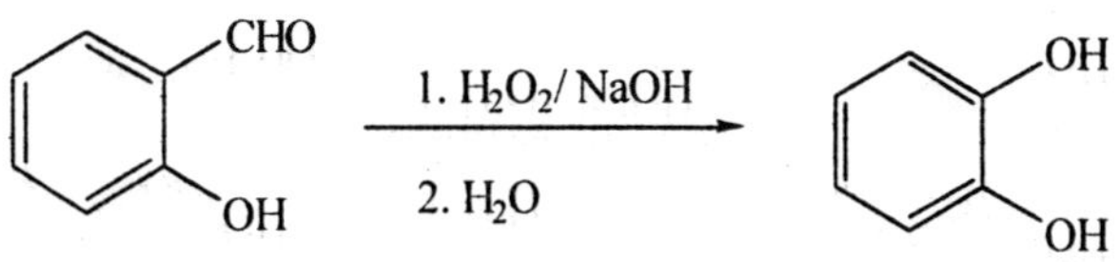

The mechanism of the reaction involves following steps

1. Attack of hydroperoxide anion on carbonyl carbon to form **55**.

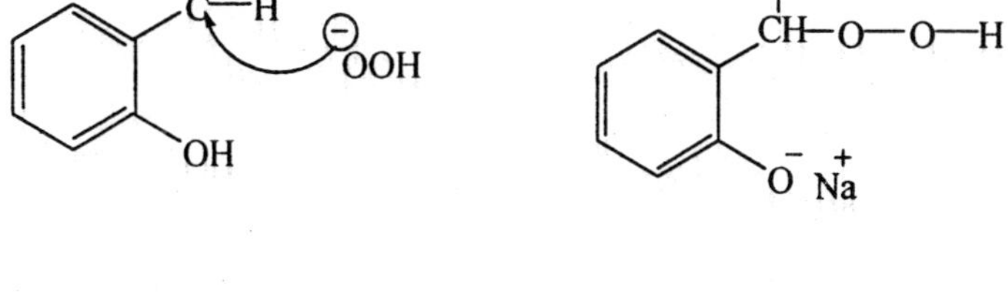

55

1. Migration of aryl group takes place with expulsion of hydroxide ion to form intermediate formate **56**.

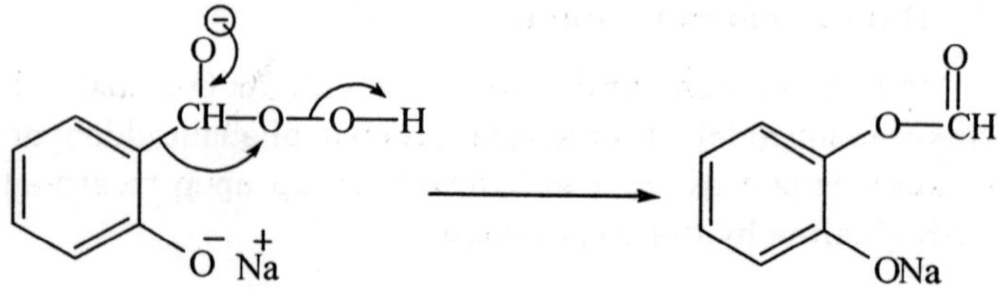

56

2. Hydrolysis of formate **56** to phenol finally takes place.

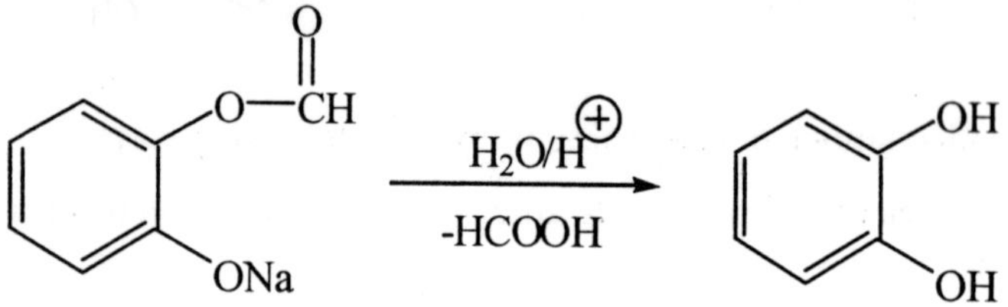

12.4. Rearrangements involving electron deficient Boron

This type of rearrangement is common in boranes. Alkyl boranes which are easily synthesized from hydroboration of alkenes are, thus, useful intermediates due to ease with which they undergo rearrangement. During the reaction a mild nucleophile having lone pair of electron is added to the empty orbital of the electron deficient boron (boron contains only six electrons). Then migration of an alkyl group to the atom which is attached to the boron, takes place. The boron is finally removed by hydrolysis.

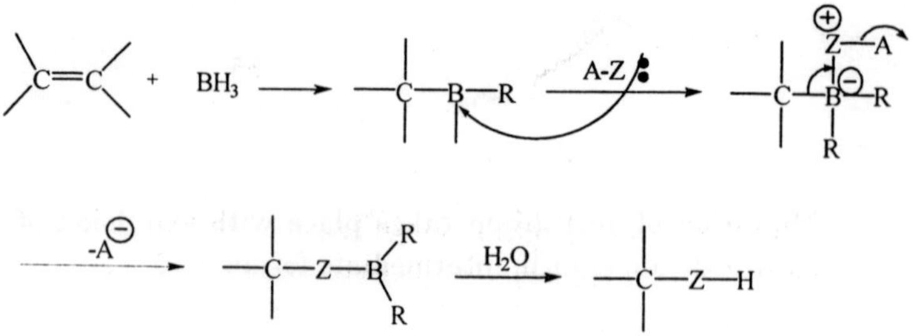

12.5. Electron rich (anionic) skeletal rearrangement

Rearrangements involving electron rich atom usually take place in the presence of a strong base. The rearrangement involves a three centered transition state which is electron rich.

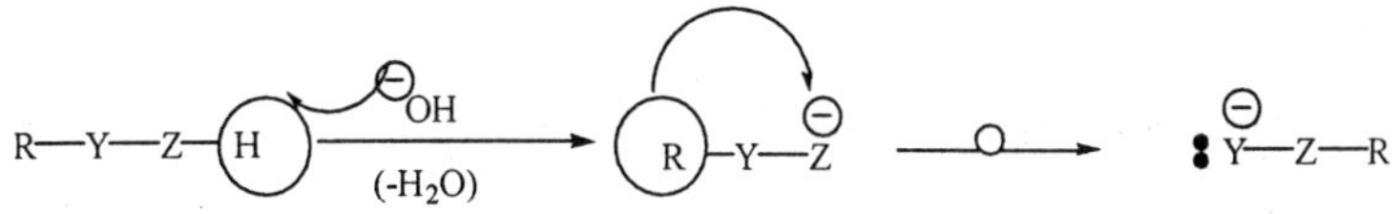

Some important examples of anionic rearrangement:

1. Steven rearrangement

The reaction involves migration of an alkyl group from a sulfonium or quaternary ammonium salt to an adjacent carbanionic centre in the presence of a base.[13,14]

The rearrangement, thus, affords a tertiary amine or sulfide.

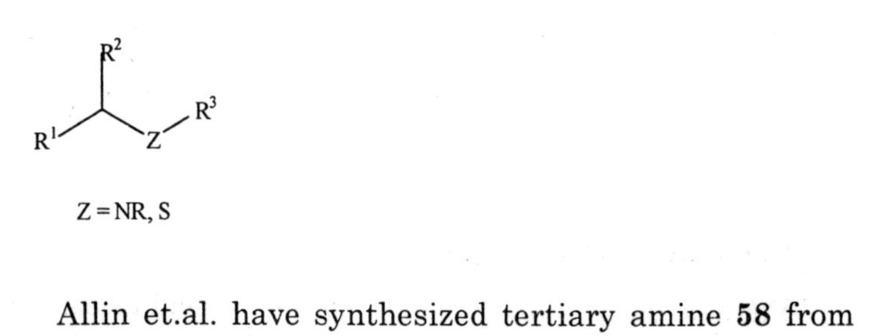

Z = NR, S

Allin et.al. have synthesized tertiary amine **58** from quaternary ammonium salt **57** in the presence of potassium tertiary butoxide in acetonitrile.[15]

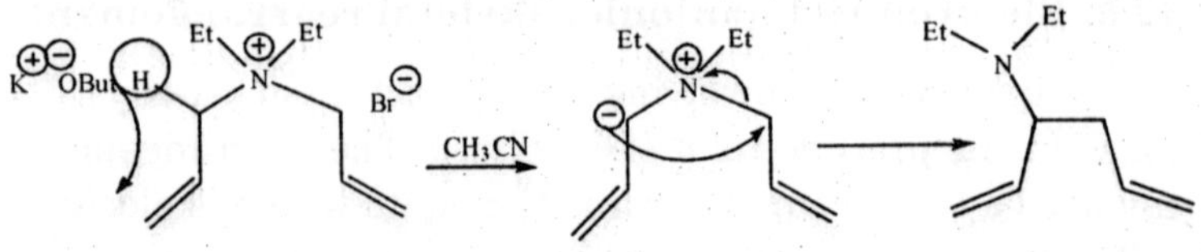

57 58

Synthetic utility of Stevens Rearrangement

1. Recently Stevens Rearrangement has been used for the synthesis of alkaloids. The reaction involves methods for generating ylide, ring expansion and strategies for stereochemical control and application to alkaloids (natural products)[16]

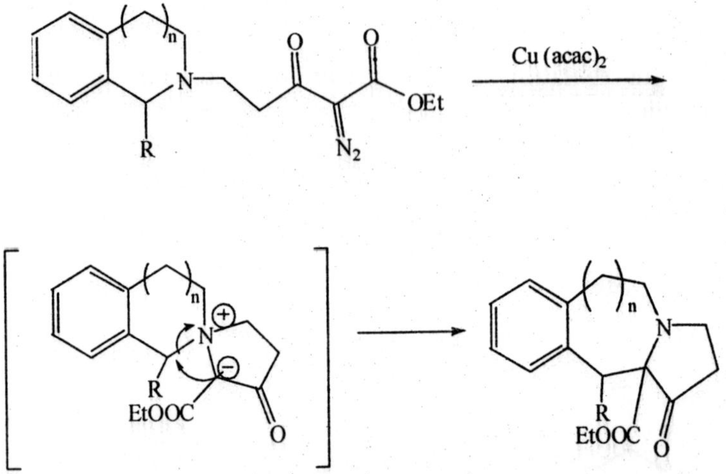

2. **Favoraskii rearrangement**

The rearrangement of α-halo ketone to acid or ester in the presence of a base is known as Favoraskii rearrangement. In case of a cyclic α-halo ketone, ring contraction takes place

12.2. Rearrangements involving migration to electron defficient nitrogen

These are the rearrangements which involve migration of a group to electron deficient nitrogen known as nitrenes.

The electron deficient nitrogen is generated by loss of a leaving group attached to nitrogen. The rearrangements involving migration of a group to nitrene follows a concerted mechanism, i.e. migration of group to nitrene and removal of leaving group from nitrogen takes place simultaneously. Migration usually involve 1,2 shift and driving force for these types of rearrangements is provided by removal of small neutral molecules such as water, nitrogen etc.

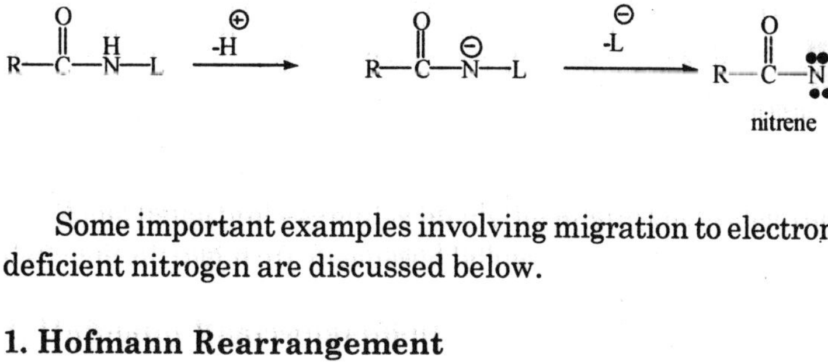

Some important examples involving migration to electron deficient nitrogen are discussed below.

1. Hofmann Rearrangement

The reaction is used for the synthesis of primary amines by treatment of primary carboxylic acid amide with bromine and aqueous or alcoholic potash (or potassium hypobromite).[5] The reaction proceeds via intermediate isothiocyanate **29**.

$$RCONH_2 \xrightarrow[Br_2]{KOH} RNH_2$$

The reaction mechanism involves following steps.

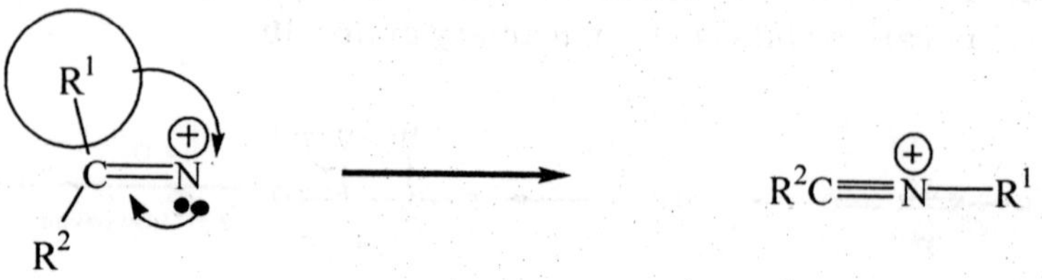

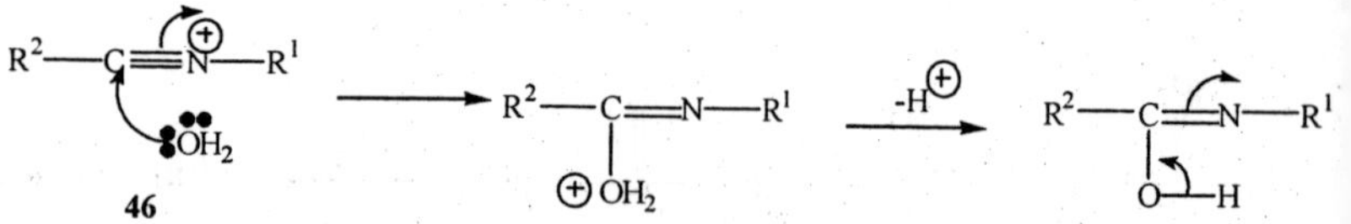

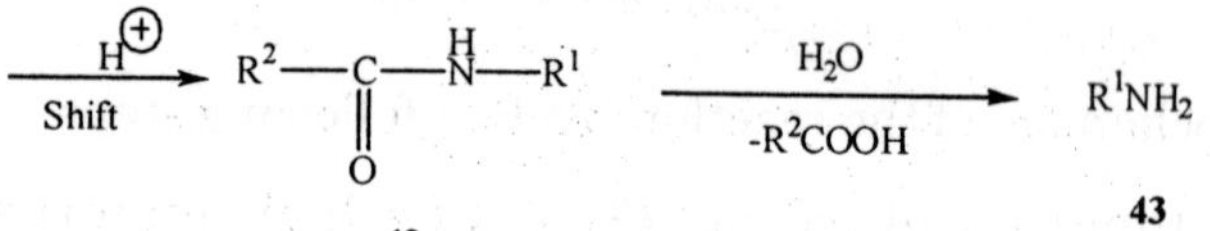

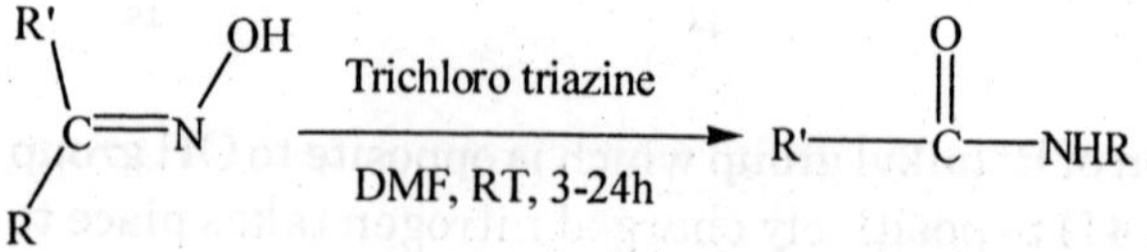

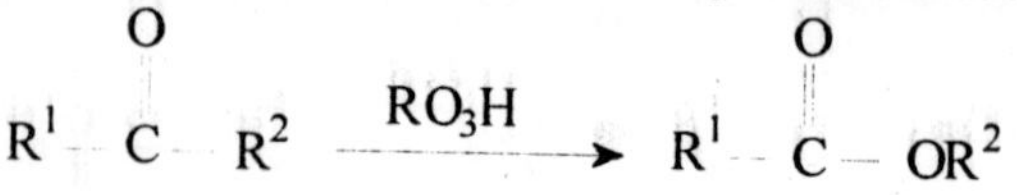

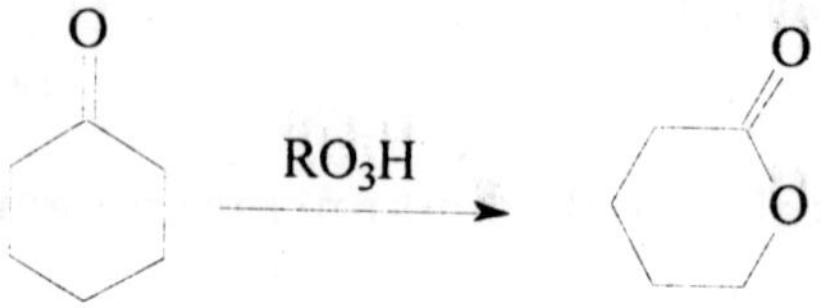

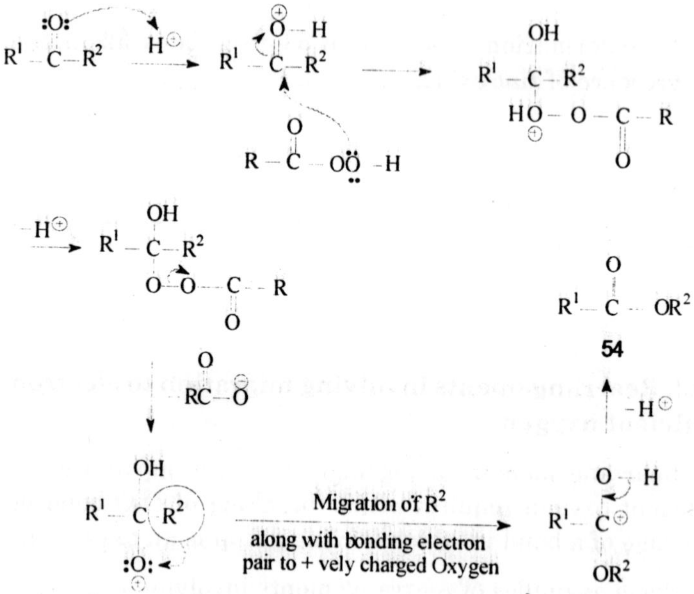

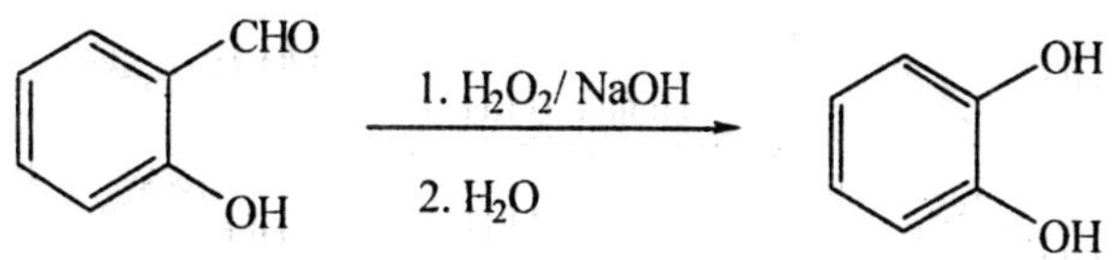

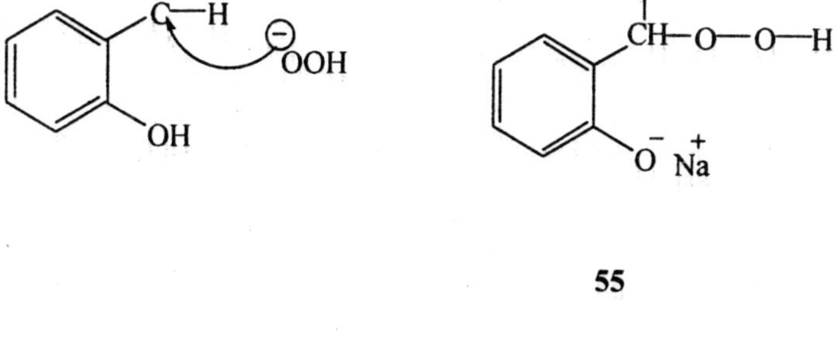

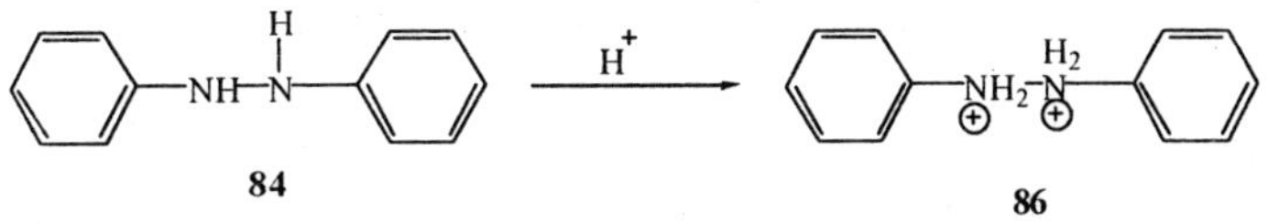

2. Breaking and formation of bonds in intermediate **86** takes place to form the product **85**.

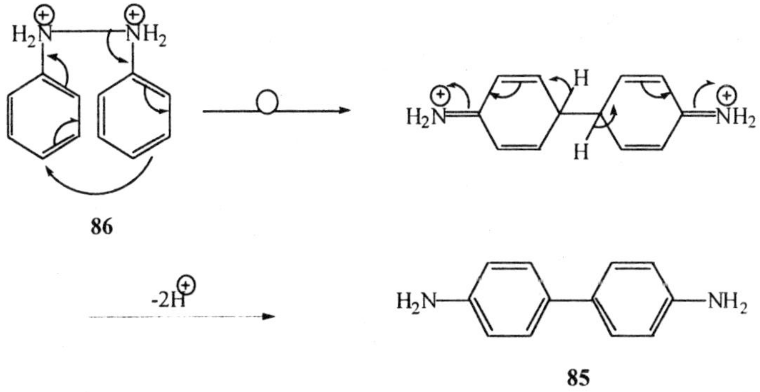

The mechanism is supported by carrying out reaction between two different hydrazobenzene when two benzidine are obtained and no cross coupled products are obtained.

Formation of various semidines is shown below.

a) Formation of o-semidine **87**

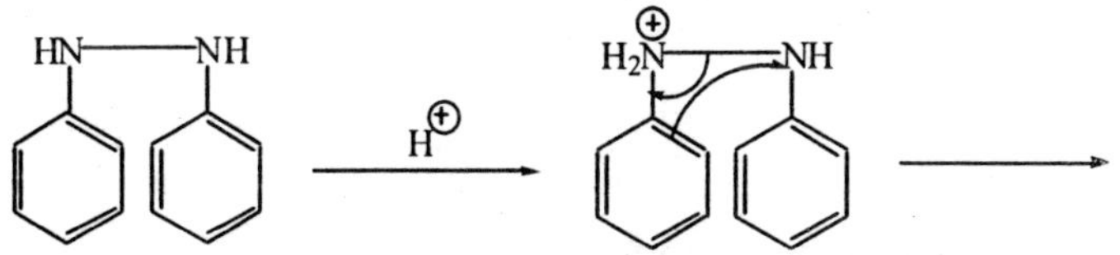

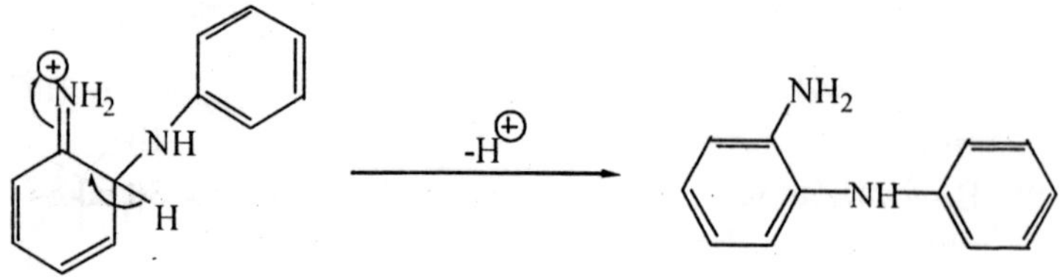

87

b) Formation of p-semidine 88

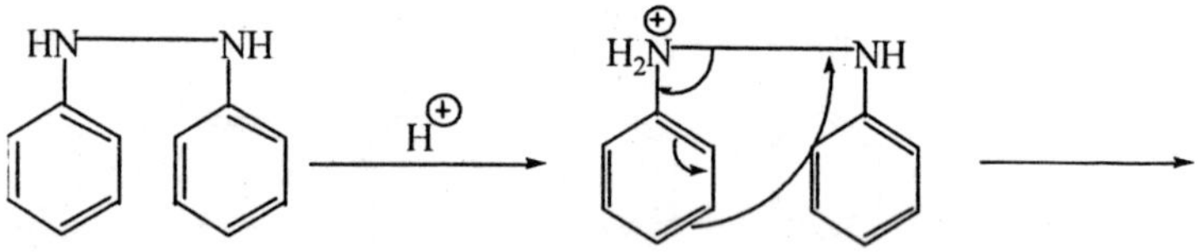

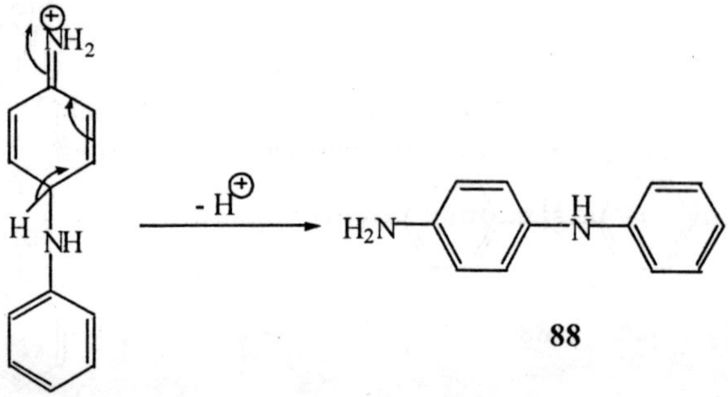

88

Formation of diphenylene **89.**

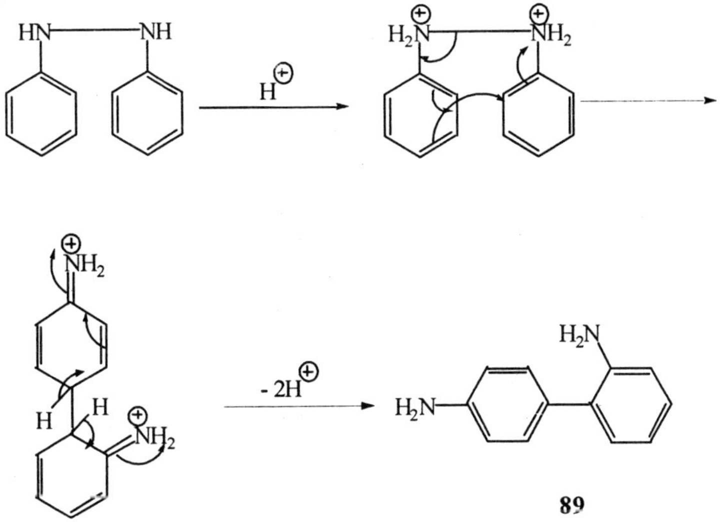

89

Effect of substituents on the formation of o/p isomer

Usually a para benzidine is obtained as major product. However, if para position is occupied by some substituent then o-benzidine/ o-semidine is obtained as major product e.g.

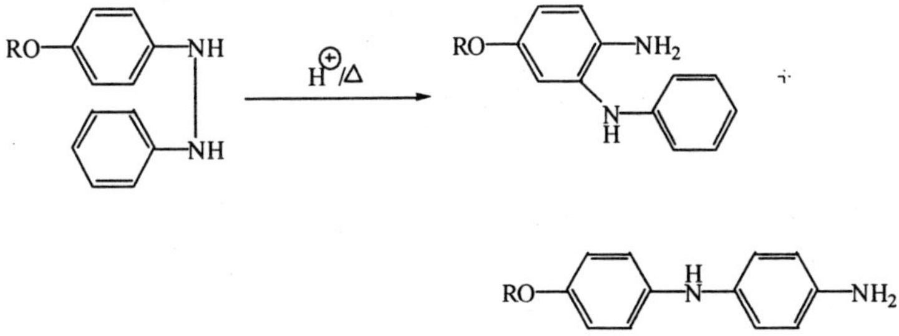

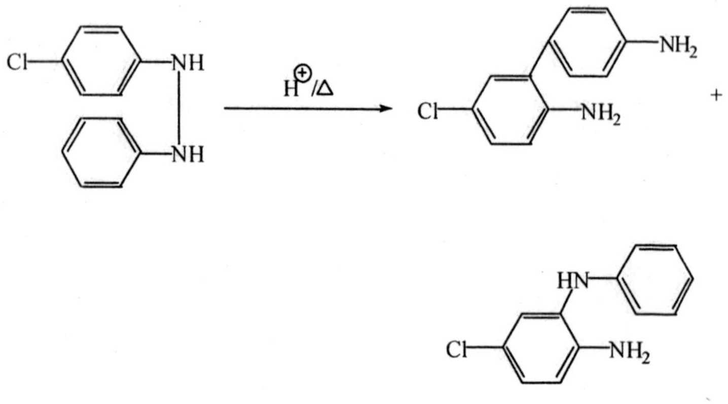

The mechanism of the formation of products is shown below:

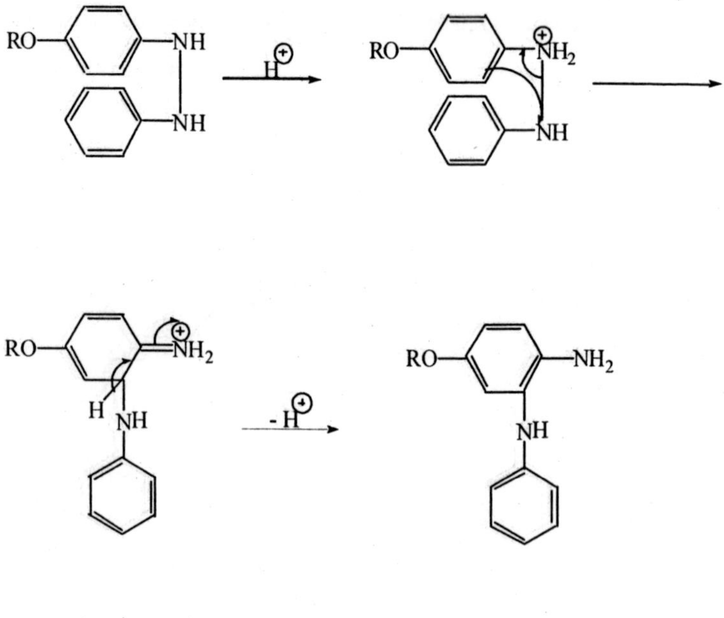

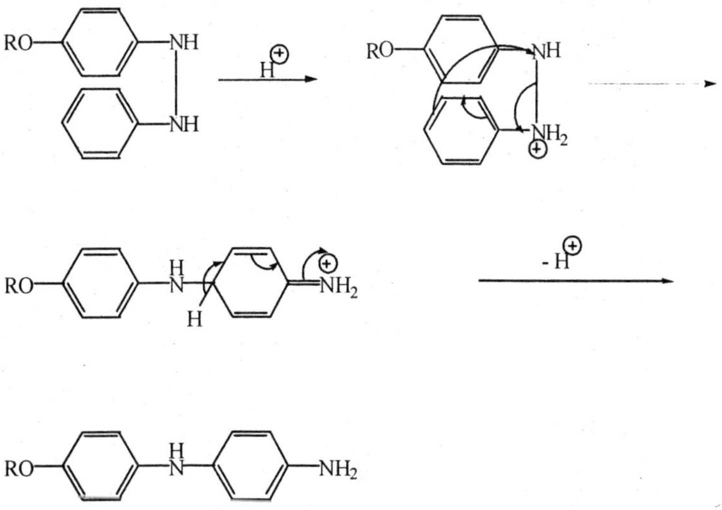

7. Hofmann-Maritus rearrangement

The reactions involves formation of ortho/para alkyl anilines from N-alkyl aniline hydrohalide upon heating at high temperature (200-300 °C)[34].The thermal rearrangement proceeds via intra molecular migration of alkyl group. For example, when N-methyl aniline hydrochloride **90** is heated strongly, methyl group migrates to ortho/para position of the aromatic ring to form o-methyl aniline **91a** and p-methyl aniline **91b**.

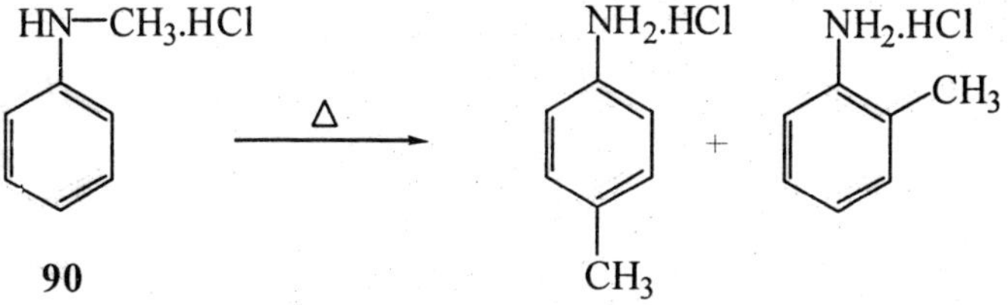

90

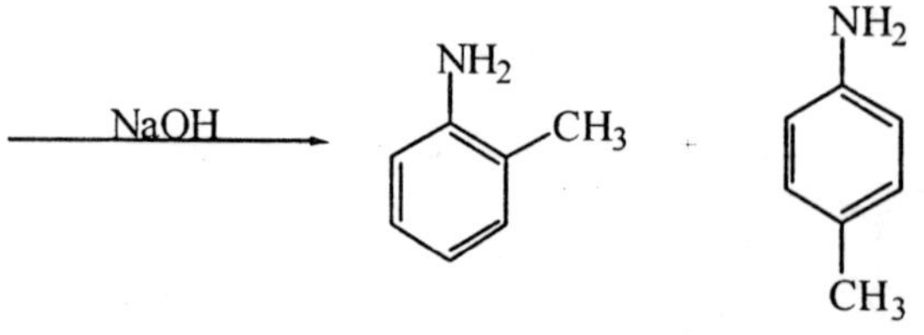

91a **91b**

If it is N, N-dialkyl aniline **92** then migration of both the alkyl groups takes place in step wise manner to form N-alkyl aniline **93** and dialkyl anline **94** respectively.

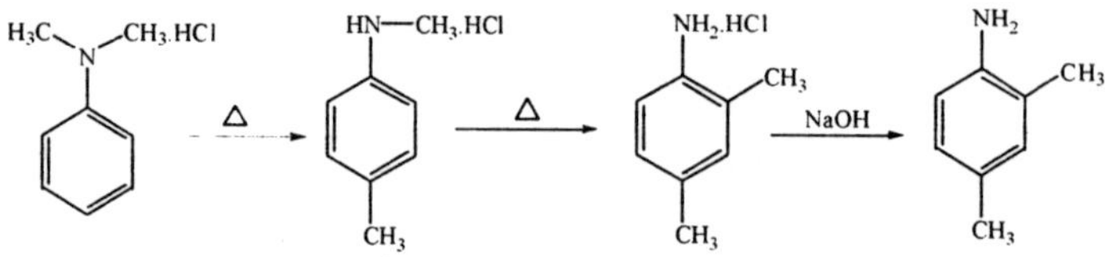

92 **93** **94**

The migration of alkyl group takes place preferentially to para position. However, if para position is occupied by some other substituent then the migration takes place exclusively to ortho position.

The rearrangement is intramolecular and involves following steps:

1. The formation of methy chloride by SN² reaction of N-alkyl anilinium hydrochloride **90** with chloride ion.

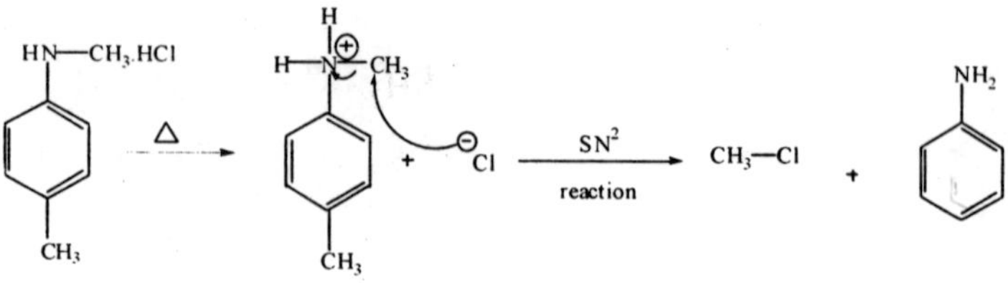

90

2. Aniline and methyl chloride react with each other through Friedel Craft alkylation reaction to form the product.

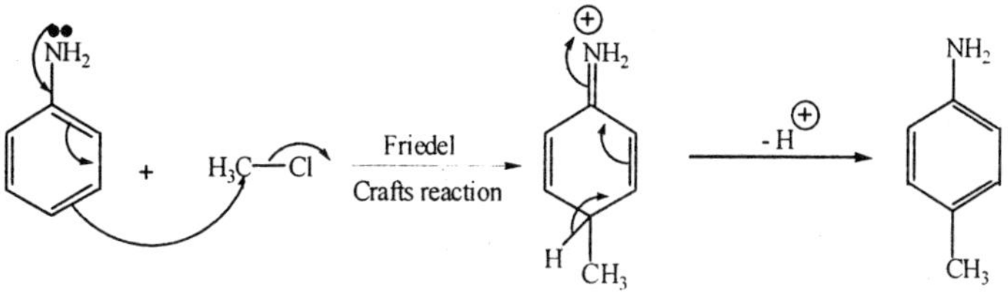

If alkyl groups are secondary or tertiary then formation of alkyl halide takes place by SN[1] reaction,e.g.

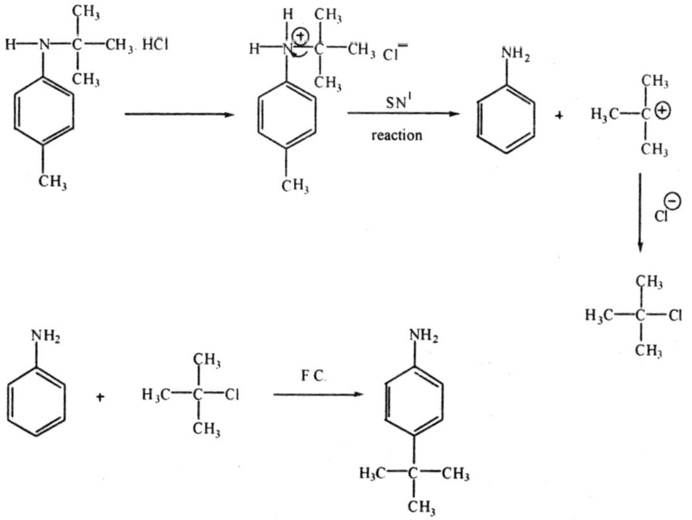

Alkyl groups which are prone to rearrangement, giver rearranged product,e.g.butyl aniline hydrochloride forms tertiary butyl aniline.

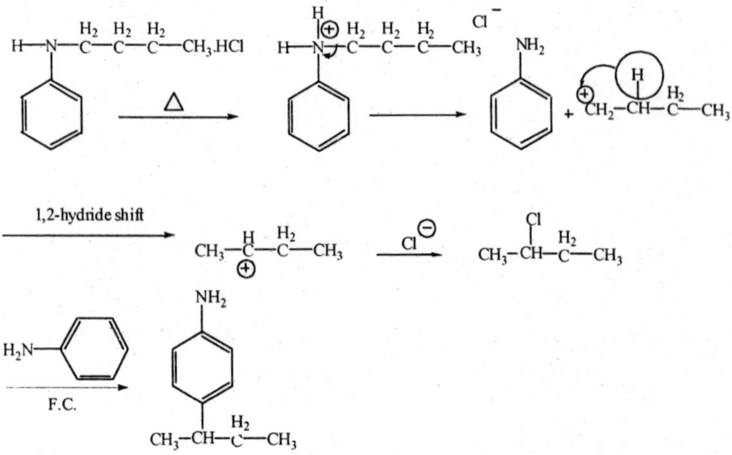

REFERENCES

1. R. Fitting, *Ann.* 1860, **54** ,114.
2. B.P. Mundy, Y. Kim and R.J. Warnet, *Heterocycles*, 1983, **20**, 1727.
3. L. Wolf, *Ann.*,1912, **394**, 25.
4. J. Liebig, *Ann.*, 1838, **25**, 27.; N. Zinin *ibid*, 1939, **31**, 329.
5. A.W. Hofmann, *Ber.*, 1881,. **14**, 725.
6. W. Lossen, *Ann.* ,1872, **161**, 347.
7. T. Curtius, *Ber.*,1890, **23**, 3023.
8. L. De Luca, G. Giacomelli and A. Porcheddu, *J. Org. Chem.*, 2002, **67**, 6272.
9. S. Chandrasekhar and K. Gopalaiah, *Tetrahedron Lett.*, 2003, **44**, 755

10. R. Ramalingum and Y. T. Park, *J. Org. Chem.* ,2007, **72,** 4536.

11. H. Sharghi and M. Hosseini, *Synthesis* ,2002, 1057.

12. A. Baeyer, V. Villiger, *Ber.* ,1899, **32**, 3625; ibid, 1900, **33**, 855.

13. T.S. Stevens, E.M. Creighton, A.B. Gordon and M. McNicol, *J. Chem. Soc.*, 1928, 3193.

14. T.S. Stevens, *J. Chem. Soc.*, 1930, 2107.

15. S. M. Allin, M. A. C. Button and S. J. Shuttleworth, *Synlett*, 1997, 725.

16. John A. Vanecko, Hayley Wan, Frederick G. West *Tetrahedron*, 2006, **62(6)**,1043.

17. G. Wittig and L. Lohmann, *Ann.* , 1942, **550**, 260.

18. M. Kunishima, K. Hioki, K. Kono, A. Kato and S. Tani, *J. Org. Chem.* ,1997**, 62**, 7542.

19. X. Han and D. W. Armstrong , *Org. Lett.*, 2005, **7**, 4205

20. K. Wang, C. J. Burgard and S. G. Nelson, *Org. Lett.*, 2007, **9**, 2325.

21. M. H. Suhre, M. Reif and S. F. Kirsch, *Org. Lett.* , 2005, **7**, 3873.

22. B. D. Sherry and F. D. Toste, *J. Am. Chem. Soc.*, 2004, **126**, 15978.

23. Y. Okada, T. Ureno. and T. Hayashi, *J. Oleo Sci.*, 2003, **52 (9),** 491.

24. K. Hiratani, K.Kasuga, M. Goto and H. Uzawa, *J. Am. Chem. Soc.*, 1997, **119**, 12677

25. G. Dimartino and J. M. Percy, *Chem. Commun.*, 2000, 2339.

26. E. Bamberger, *Ber.* , **27**, 1347, 1548, 1894.

27. K. Von Auwers and K. Ziegler, *Ann.*, 1921, **425**, 217.

28. O. Wallach and L. Belli, *Ber.*, 1880, **13**, 525.

29. J. Yamamoto, Y. Nishigaki, M. Umezer and T. Matsuura, *Tetrahedron,* 1980, **36**, 3177.

30. J. Sing, P. Singh, J. L. Boivin and P. E. Gagnon, *Canadian J. Chem.*, **41**, 1963, 499.

31. A. Lalitha, K. Pitchumani and C. Srinivasan, *J. Molecular Cata. A.:Chem.*, 2000,**160** (2), 429.

32. K. Fries and G. Fink,*Ber.,* 1908, **4**,4271;K. Fries and Pfaffendorf, *ibid* , 1910,**43**,212.

33. A. W. Hofmann,. *Proc. Roy. Soc London*, 1863, **12**, 576.; P. Jacobson et. Al., *Ber.*, 1893, **26**, 688.

34. A. W. Hofmann and C. A. Maritus, *Ber.*, 1871, **4**, 742.

35. R.E. Ireland and R.H. Mueller, *J. Am. Chem. Soc.,* **94**, 5897.

36. A.E. Wick, K. Steen, A. Eschenmoser, *Helv. Chim. Acta,* 1964, **47**, 2425.

37. W.S. Johnson et.al., *J. Am. Chem. Soc.,* 1970, **92**, 741.

Bibliography

Ahluwalia, V.K. and Parashar, R.K., *Organic Reaction Mechanism*, Narosa Publishing House Pvt. Ltd, 2007.

Finar I.L., Organic Chemistry, Vol. I & II, *The Fundamental Principles*, The English Language Book Society and Longman Group Ltd., 1973.

Jain, M.K., *Organic Chemistry for Degree Students*, Shoban Lal Nagin Chand & Co.

Mayo, P. De, *Molecular Rearrangement, Vol.II*, Interscience Publishers, 1964.

Morrison, R.H. and Boyd, R.N., *Organic Chemistry*, Prentice Hall of India Pvt. Ltd., New Delhi, 1978.

Arun Bahl and B. S. Bahl, *A Textbook of Organic Chemistry*, S. Chand and Company Ltd., New Delhi, 2005.

P. L. Soni, and H. M. Chawla, *A Textbook of Organic Chemistry*, Sultan Chand and Sons, New Delhi, 1996.

Jerry March, *Advanced Organic Chemistry*, John Wiley and Sons, New York, 2005.